The Flower Mound
Flower Mound, Texas

A History and Field Guide to the Flowers and Grasses

Alton Bowman
The Mound Foundation

THE
DONNING COMPANY
PUBLISHERS

Title Page: View of the Flower Mound sign on the south fence frontage, looking north from FM 3040.

All photographs are by the author unless otherwise credited. All were taken with a Nikon F 100 on slide film. The lenses were Nikkor 60 mm Micro and 24–50 Macro Zoom.

To my wife, Sweety, who introduced me to the Mound and the rewards of volunteerism. Her loyal support made this book possible.

Copyright © 2004 by The Mound Corporation DBA The Mound Foundation

All rights reserved, including the right to reproduce this work in any form whatsoever without permission in writing from the publisher, except for brief passages in connection with a review.
For information, please write:

The Donning Company Publishers
184 Business Park Drive, Suite 206
Virginia Beach, VA 23462

Steve Mull, *General Manager*
Barbara B. Buchanan, *Office Manager*
Pamela Koch, *Editor*
Stephanie Danko, *Graphic Designer*
Amanda Guilmain, *Imaging Artist*
Mary Ellen Wheeler, *Proofreader*
Anne Cordray, *Project Research Coordinator*
Scott Rule, *Director of Marketing*
Travis Gallup, *Marketing Coordinator*

Barbara Bolton, Project Director

Library of Congress Cataloging-in-Publication Data

Bowman, Alton, 1945-
 The Flower Mound, Flower Mound, Texas : a history and field guide to the flowers and grasses / Alton Bowman.
   p. cm.
 Includes bibliographical references and index.
 ISBN 1-57864-290-6 (softcover : alk. paper)
 1. Angiosperms--Texas--Flower Mound--Identification. 2. Grasses--Texas--Flower Mound--Identification.
 3. Flower Mound (Tex.)--History. 4. Flower Mound (Tex.)--Guidebooks I. Title.
 QK188.B69 2004
 582.13'09764'555--dc22
                              2004022302

Printed in the United States of America at Walsworth Publishing Company

Table of Contents

4   *Foreword: Bob Rheudasil*

7   *Preface*

8   *Contributors*
                 Honoring and Remembering

9   *Introduction*

10   Chapter 1 *History*
                 Geologic Description • The Black Land Prairie • The Cross Timbers
                 Prehistory • Native American Archaeology • Modern Native American History
                 Indigenous Wildlife • Bison • European Contact • First American Settlers • The Prairie and the Settler

18   Chapter 2 *Ownership of the Mound*
                 John and Edy Wiswell • George and Hattie Beavers • Ray and Charlie Fay Lester
                 Edward and Betty Marcus: Black Mark Farms; Stanley Marcus on Edward Marcus
                 Flower Mound New Town • Bellamah Community Development

24   Chapter 3 *Related History*
                 Religious Services on the Mound • Good Neighbors: The Donald Community;
                 The Flower Mound Presbyterian Church; The Alexanders • Mound Legends and Lore

28   Chapter 4 *The Mound Corporation*
                 Otto Consolvo, Founding Chairman: The First Fence; The Historic Marker; The Monument
                 Ronnie Hilliard Chairmanship: The Wrought Iron Fence • Bill Neiman Chairmanship: The Prairie Picnics
                 Alton Bowman Chairmanship: Fence Completion; Documentation and Photography of the Plants
                 Sale of the Frontage for 3040 Expansion; Archaeology Update; Gilgai (Hog Wallows); Remaining Wildlife on the Mound

34   **Field Guide**
                 How to Use the Field Guide

35   Chapter 5 *The Awakening*
                 The Flowers of February and March • Color Chart

53   Chapter 6 *Explosion of Color*
                 The Flowers of April and May • Color Chart

88   Chapter 7 *Drought Tolerant*
                 The Flowers of June through November • Color Chart

100   Chapter 8 *The Majestic Grasses*

107   Chapter 9 *The Dormant Season*

108   *Mound Species Identified but Not Reproduced*

108   *Bibliography*

110   *Index*

112   *About the Author*

Foreword

by Bob Rheudasil

It was the spring of 1953 when I first saw the Mound. I had heard about it many times before but could only imagine what its namesake must look like until Mr. Edward Marcus drove me past it. I had come to Flower Mound to discuss a job with Mr. Marcus. He had a dream to raise prize Angus cattle on a 3,500-acre farm named the Black Mark Farms and wanted me to help him make it come true.

I came from east of Paris, Texas, in Lamar County. All around where I grew up, there was a lot of prairie grass, and it interested me. In fact, "The King of the World" in prairie grass is located there. I knew how special the Mound was right away.

In 1957, I bought a 320-acre parcel to add to Black Mark Farms for Mr. Marcus that included the Mound in the purchase. I wanted to buy it for $350 an acre, but Mr. Marcus settled for $375 instead. It proved to be a bargain and great source of pride for him through the next thirty-five years.

At the time Mr. Marcus purchased the Mound, FM 3040 was just a gravel country road and FM 2499 had just been completed. Back then, and for many, many years, the only lights you could see from the Mound were in downtown Dallas and nowhere else. In those days, there was nothing but flatland prairie surrounding the Mound for some distance. The Mound rose gently above the prairie and was a knee-deep lush meadow of prairie grass.

Knowing it was the only native prairie grass anywhere around here, we never even thought to disturb it or plow it up. Once you plow up prairie grass, it is gone forever! I planted a few trees and bluebonnets around the base of it in the late 1950s, but otherwise, except for baling hay every August to feed our horses and some cattle, we left it and let nature take care of it. Little did I know at the time that I would see the Mound through more than fifty seasons.

Visitors to Black Mark Farms and Flower Mound always wanted to see the Mound. Many were famous, but none was more interesting and kind than the former First Lady, Lady Bird Johnson. She had an extensive knowledge of flowers and plants. So, when Mr. Marcus, the Secret Service, and I took her up on the Mound, she could identify every plant but one. After she left, I got that flower identified for her and sent her the information on it. I believe it was Lambert's Locoweed. It was the early 1970s, and at that particular time on the Mound, looking east, there were no houses and

Edward Marcus and Bob Rheudasil (just off camera, right) host Lady Bird Johnson on a visit to the Mound in 1970. The Mound was a part of Flower Mound New Town at this time, and Mrs. Johnson was studying beautification plans for the city.

just acres and acres of crepe myrtles in full bloom. I remember she said it was the most beautiful sight she had seen in a long, long time.

Mr. Marcus and I took countless people up on the Mound. Everyone would always remark how incredible the view was. Eventually, you could see the glow of the lights of Fort Worth. And after the Texas Woman's University dorms were built, we could even see those lights. Years and years ago, there must have been prairie grass all over this country. But in our area, that is now all that remains. I am glad it was in the hands of people like Mr. Marcus, who kept it intact and was careful not to destroy or develop it.

The Mound was also instrumental for educational purposes. For instance, in the 1950s, the Corps of Engineers used it to train their Rangers to identify various species of native prairie grasses. They sent Rangers from different parts of the state out of the Fort Worth office to spend time on the Mound during construction of both Lake Lewisville and Grapevine Lake.

Of course, it was also a source of inspiration. As early as the 1950s, the first organized event I recall on the Mound was attended by the Denton County Ministerial Alliance. In fact, Channel 4 covered the event, which included a group of ministers touring with the Soil Conservation Service, who came to appreciate the beauty of the land. Many others have come since then to enjoy the view and peace that still remains among those native grasses. No one ever asked me to get married on it. And they probably should have, as it was not designated as public land when Mr. Marcus owned it and was always posted accordingly. But I think some people did choose that site to be married, and you can surely understand why they would.

Edward Marcus, left, an unidentified local minister, center, and a Denton County Agricultural Extension agent meet on the Mound to study wild foxglove in the 1950s. Photo courtesy of Black Mark Farms.

All kinds of tales about the Mound have been told, including the notion that it is a sacred Indian burial ground. I can't personally substantiate any of those claims. What I do know is that Mr. Marcus asked members of the Southern Methodist University Archaeology Department to dig various areas on the Mound, and they found no evidence of any burial grounds or Indians. But, nevertheless, it is an extraordinary place.

Mr. Marcus always said it was the perfect place for a sunrise service, and I remember he was just thrilled to attend the first one ever held on the Mound. Each year, when the preparations are being made for those events, I can't help but think how proud he would be that the traditions endure and the Mound continues to be appreciated.

In that regard, I would like to say that I couldn't imagine anyone working as hard or as diligently as the Mound Foundation Board. Its members have done an outstanding job, and I hope their work will continue for generations and generations to come. Alton Bowman has served as the head of the Mound Foundation Board, and no one could have put more time, interest, or foresight into preserving this incredible land.

I extend my deepest gratitude to the Mound Foundation, Alton Bowman, and everyone throughout the years who has worked to preserve the Mound. Because of their tireless efforts, future generations will be able to enjoy the Mound as I have. It has been a privilege to be a part of the history this land has so intricately woven in our town, and it is only right that this special namesake remain undeveloped to welcome those yet to come.

Bob Rheudasil
First Mayor of Flower Mound

The Flower Mound. Photo courtesy of Bob Rheudasil, May 1973.

Preface

My involvement with the Flower Mound began when I joined the Mound Foundation, then known as the Mound Corporation, in 1987. Only four years earlier, the Mound had been secured as a perpetual site, and it was under the competent chairmanship of Bill Neiman. Whenever Bill and I were at the Mound, I was always impressed by how many of these strange and beautiful plants he was familiar with by name. It was obvious that there was a great deal to learn from being associated with this little twelve-acre tract with the charming name.

After ten years of fundraising, fence building, and red tape acquiring a nonprofit tax status, we finally were able to begin our real work—documenting the plants. Edith Bettinger, an early member, had compiled a small brochure with line drawings of prominent plants. I began by taking slides of the blooms. The next spring, I started very early in the season and photographed everything I saw on a weekly basis. Most of the plants were new to me. I purchased all the different field guides to help identify the exact species. About this time, the Mound came under the umbrella of the Texas Nature Conservancy. This gave us access to the considerable expertise of its native prairie expert, Jim Eidson. Jim conducted a basic survey from which we started our species list.

The next big discovery was the Botanical Research Institute of Texas (BRIT) in Fort Worth, which made available to us the largest botanical collection and reference library in the Southwest. Barney Lipscomb of BRIT did a site inspection and was always available to answer questions and find anything in the extensive library. When BRIT published its magnificent *Illustrated Flora of North Central Texas*, we had a complete reference of every plant in our area.

After five years of photography and documentation, one hundred plants were listed and identified. Jack Neal of Flower Mound volunteered to assist in the book project. Jack dedicated a year of regular visits to digitally photograph, identify, and confirm or challenge my list. We used the BRIT's text as the reference to settle all conflicts. Jack wrote most of the botanical descriptions and added fifteen or more new plants to the list. His assistance allowed me to add the history section to the book. The Foundation contracted Steve Vick of Denton to research the title history of the Wiswell Survey, the original Peters Colony tract that contained the Mound. I interviewed Bob Rheudasil, the first mayor of Flower Mound and manager of Edward Marcus' holdings in Denton County. Bob provided many of the photographs and information from this period. Since many of the people involved in the study were deceased, the *Flower Mound History* was used for photos and data. A. C. Greene's study was very helpful, as was Marty Matyas' *History of Flower Mound New Town*. Archaeologists Paul and Jan Lorrain provided archaeological data. Grassman John Snowden of Bluestem Nurseries visited the Mound twice and helped to identify the grasses. Nicole Dogan provided the maps and modern aerial photography. The support of the rest of the board members was essential, and their fundraising efforts have made the printing possible. The administration of Mayor Lori DeLuca of the Town of Flower Mound provided the first financial support received from the town, and over the last five years, those funds were used to seed other fundraising projects.

This book fulfills the basic bylaws of the Foundation of historic, educational, and scientific purposes. In addition, 100 percent of the funds raised from the sale of the books will be used for perpetual care of the Mound. My hope is that it will inspire people to visit the Mound more often to find and identify the flowers illustrated here and to answer some of the frequently asked questions about the Mound. The Mound is a gift for all to enjoy. This book is my gift to the Mound.

Alton Bowman
Flower Mound, 2004

Contributors

Florence and Bob McCarley
Stewart Title of North Texas
Flower Mound Rotary Club
Carol Norman McCarty
Lake Forest Homeowners Association
The Duke Family
Rick and Terry Lust Family
Tom Thumb
Mr. and Mrs. Al Picardi
Denton Area Teachers' Credit Union
Cyndy Dwyer
Leland and Mary Mebine
Marie Rushing-Slayton
UPS Store #3322, Flower Mound

The Mound Foundation Board of Trustees at time of publication, 2004. Left to right, Bruce Thompson; Lori Wilkins, vice chairman; Nelson Ringmacher; Albert Picardi; Andrew Eads; Joan Nelson, treasurer; Scott Hartwell; JoAnn Hanson, secretary; and Alton Bowman, chairman.

Honoring and Remembering

The Hanson Family, honoring grandmother Grace Barker, a prairie pioneer mother
John Immel, former Mound trustee, mayor, and city councilman
John Nelson, former Mound Foundation treasurer
Otto Consolvo, Mound pioneer
Andre Gerault, Mound member
Ernest Hilliard, Mound member

The Mound in a spring thunderstorm in 2004.

Introduction

The Flower Mound is the namesake of Flower Mound, Texas, located in North Central Texas, southern Denton County, and thirty miles north of Dallas-Fort Worth. The Mound is at the confluence of the Great Black Land Prairie and the Eastern Cross Timbers bioregions. It is six hundred and sixty-two feet above sea level and rises some fifty feet above the surrounding community. Nineteenth century settlers named it for the profusion of wildflowers that grew there every spring. By strokes of luck and wisdom, the Mound was saved from the plow and developers and in 1983 was sold for ten dollars to the citizens of Flower Mound to be set aside forever as a native prairie preserve. The following is a history of these events and a documentation of some of the plants that survive there.

The Mound is located on the northeast corner of the intersection of FM 3040 (Flower Mound Road) and FM 2499 (Long Prairie Road), just east of the Tom Thumb grocery store. Map courtesy of Nicole Dogan.

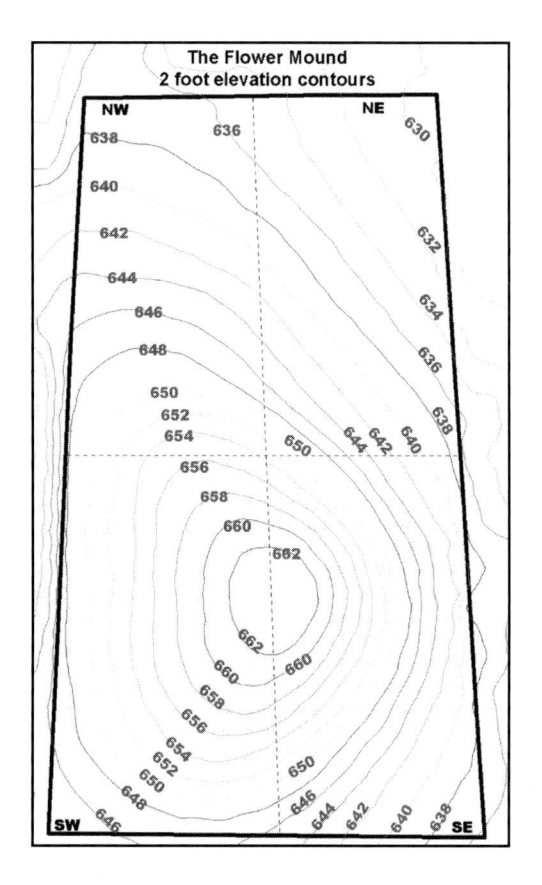

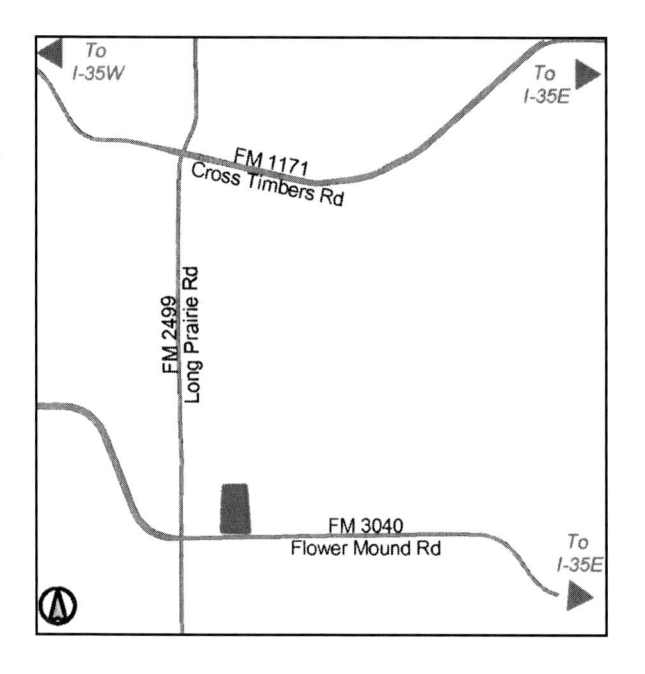

Topographical map of the Flower Mound property. Map courtesy of Nicole Dogan.

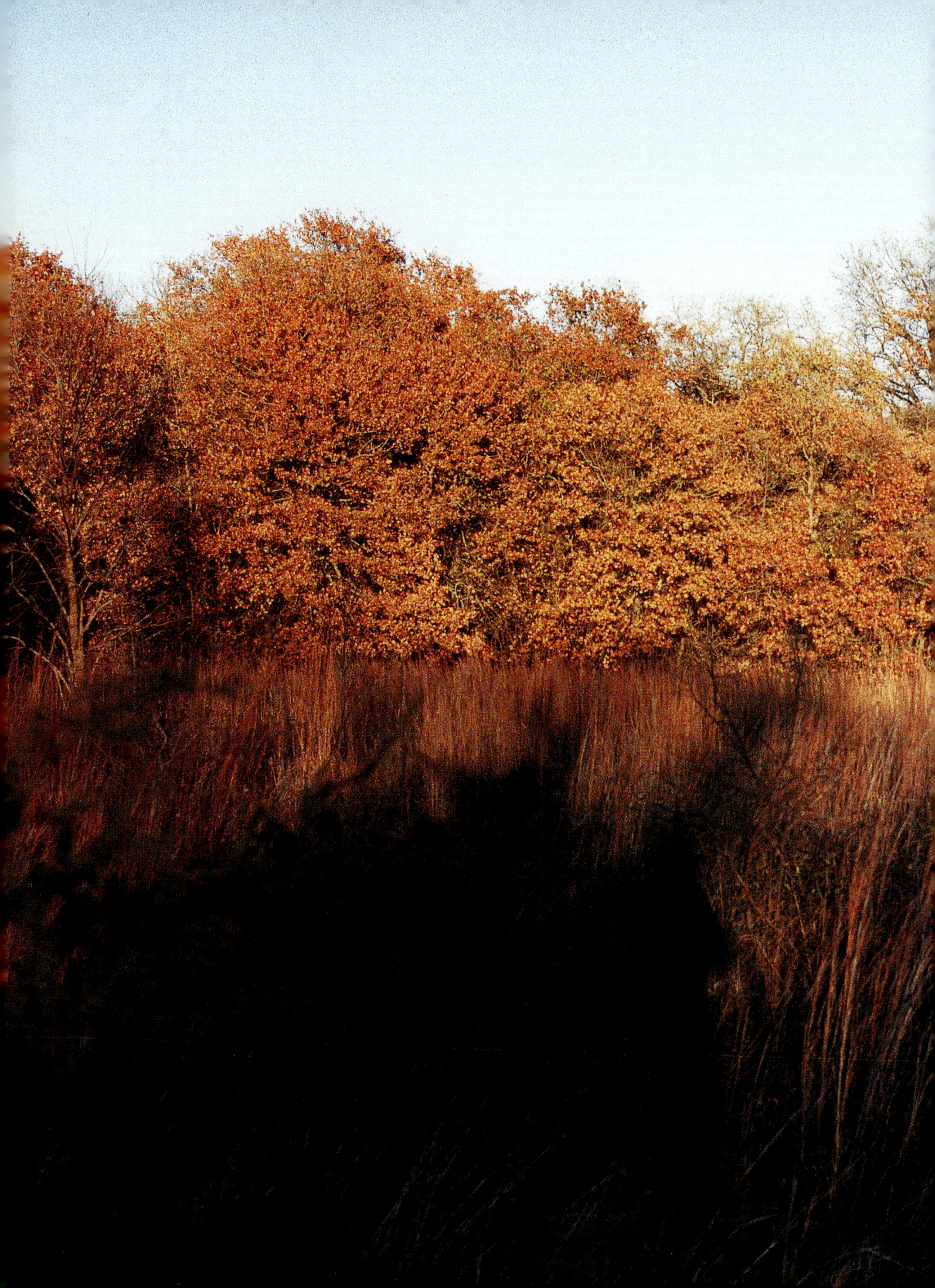

Chapter 1
History

Geologic Description

The Mound was defined in the Cretaceous Age, 65 to 145 million years ago, when a vast inland sea completely covered Texas and connected the Gulf of Mexico with the Arctic Ocean. Fossilized limestone, sands, and clays were deposited over millions of years. After the sea receded, gradual erosion and decomposition of these limestones eventually created the Black Land Prairie. The remaining sand deposits created a perfect situation for the post oak forest, locally called the Cross Timbers. The area is known geologically as the Woodbine Formation. It is the last fragment of the Great Eastern Forest to the east and the beginning of the Grand Prairie to the west.

Above: A cross-section of the Mound cut during grading for the FM 3040 expansion, showing the rich black land clays of the native prairie under the old roadbed.

Opposite Page: Fall foliage shows a typical view of the Eastern Cross Timbers post oak forest adjacent to remnant bluestem prairie in west Flower Mound.

The Black Land Prairie

The Texas Black Land Prairie runs from the Red River near Sherman east to the East Texas forests and south through Dallas and Austin to San Antonio. It covers roughly ten million acres of land and is the richest and most productive land in Texas. It is part of the Great American Prairie that extends north all the way to Canada, east to the Mississippi River, and west to the Rocky Mountains.

This tall grass prairie ecosystem has been in place for approximately ten thousand years. The main grasses are Big Bluestem, Small Bluestem, Indian Grass, Switch Grass, and the Texas state grass, Sideoats Grama. Cohabiting with the grasses is an incredible diversity of wildflowers. A healthy prairie can have two or three hundred species, depending on size and variety of terrain. The prairie was created during a period of dryness, and the plants developed deep root systems. Since a large part of the plant was underground, it survived droughts, fire, and repeated grazing by huge herds of bison. Plowing and farming destroyed the majority of this great grassland. Current estimates state that only as little as 1/2 of 1 percent of the original prairie remains. The North American Black Land Prairie has been listed with the Convention on International Trade of Endangered Species in Appendix I, eminently threatened with extinction.

The Cross Timbers

The Flower Mound lies in the middle of the Eastern Cross Timbers, one of the ten vegetation areas of Texas. The Cross Timbers is a predominantly post oak-blackjack forest, the last vestige of the eastern woodlands before opening westward onto the Great Plains. Other important tree species are bur oak, walnut, hackberry, juniper, cedar elm, Osage orange, ash, and black willow. The Cross Timbers runs from southern Kansas through Oklahoma and in Texas from the Red River to Waco. It is a straight, narrow band from fifteen to fifty miles wide, bordered on either side by prairies. Near the Flower Mound, post oaks

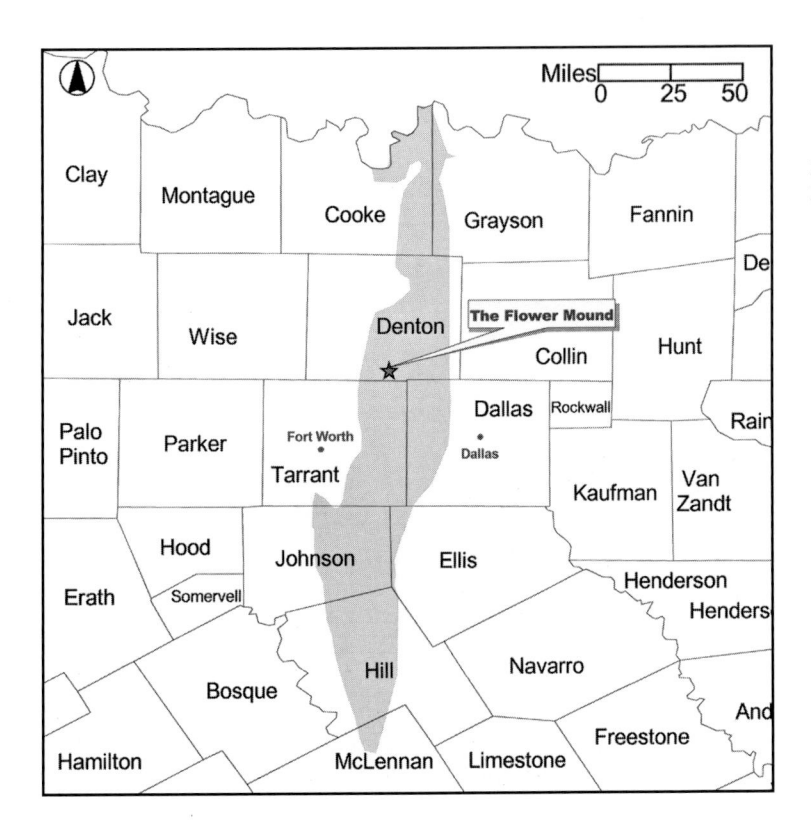

The Flower Mound, Flower Mound, Texas: A History and Field Guide to the Flowers and Grasses

Map locating the Mound in the Eastern Cross Timbers. Map courtesy Nicole Dogan.

Map locating the Mound and the Eastern Cross Timbers in the Black Land Prairie. Map courtesy of Nicole Dogan.

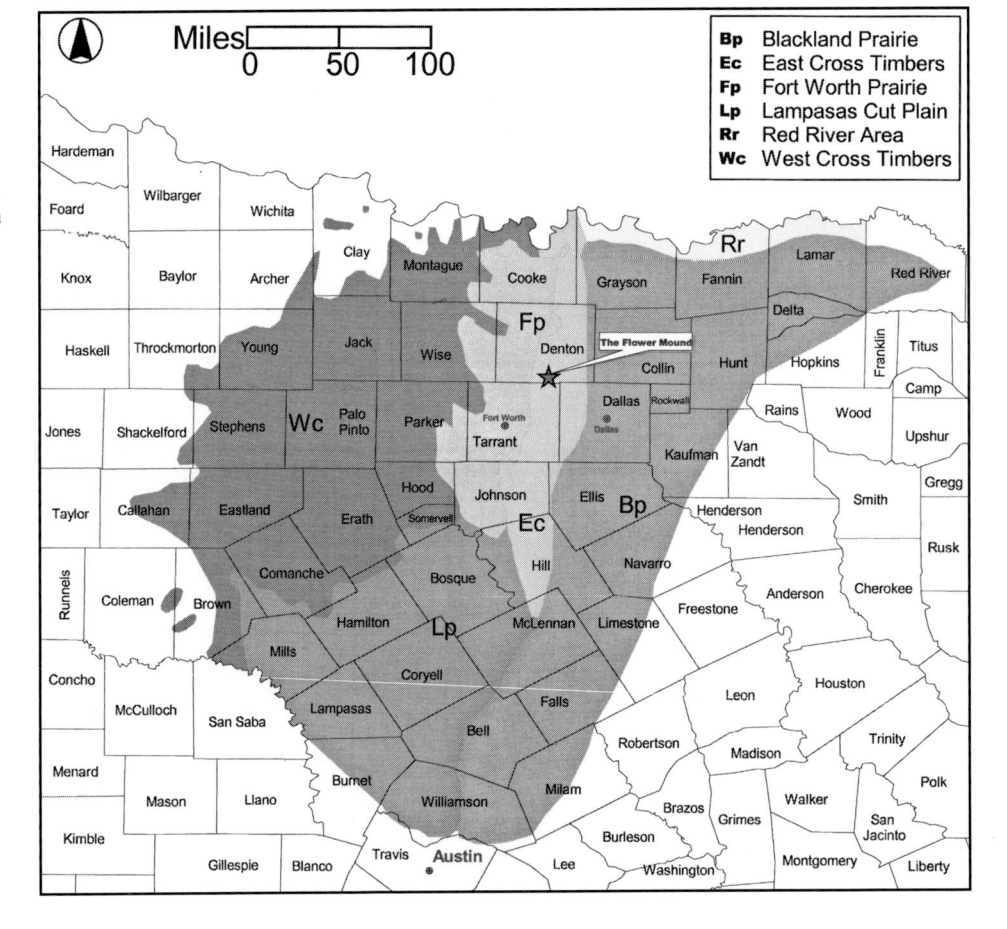

Chapter 1 History

Left: A typical specimen of old growth post oaks that may be 200 to 250 years old. This one is on Shiloh Road and shows the characteristic signs of age in gnarled growth and limb loss.

The Staten Oak, a historic double post oak on Morris Road, is a preservation success story. A drip line perimeter fence installed by the developer protects it. Insert is the tree in its natural state before development.

can be found that are 250–300 years of age. In the Oklahoma Cross Timbers, specimens have been found 350–400 years of age. The Cross Timbers is the largest discovered section of virgin forests remaining in the Eastern United States. In Flower Mound's Cross Timbers stand two of the largest bur oaks in Denton County and the recently dethroned state champion green ash. The largest known post oak in Flower Mound is forty-eight inches in diameter and is estimated at 300 years of age.

Prehistory

Many fossils of dinosaur tracks have been found near the Flower Mound, as at the Grapevine Lake spillway after the 1981 flood. A complete mosasaur was found during the excavation of D.F.W. Airport and is mounted in the airport. Recently, one mile south of the Mound, on a roadside cut for the 2499 extension, was found the oldest hadrosaur head. A duck-billed dinosaur, the fossil is now at the Dallas Museum of Natural History. Large mammoth and mastodon fossils have been found in Grayson and Dallas Counties. At the eastern edge of Denton County, excavations have revealed fossils of giant bison, lions, a species of elephant, and even camels!

A cast of the dinosaur tracks found beneath the spillway of Grapevine Dam after being exposed by the 1981 flood.

Native American Archaeology

Indian lore and the Mound are inextricably entwined. The stories and legends passed by Indians to settlers were instrumental in the ultimate preservation of the Mound as a sacred site. One of the earliest sites with evidence of humans in North America is at nearby Lake Lewisville. If carbon dating is correct, humans were here forty thousand years ago. Also, at nearby Aubrey, is an ancient bison hunt area with Clovis era points, around ten thousand years old. Three miles south of the Mound, where Denton Creek crosses 2499, ancient campfires and mussel shells have been excavated. This area has been studied since the 1940s and has yielded many lithic points, manos, drills, and preforms from the Archaic period, 2000 B.C. to A.D. 700. Archaeologists determined the Lower Denton Creek area was visited frequently prior to A.D. 700, probably in the fall to exploit mussels and nuts.

Native American atlatl point found on Redbud Point by Stanley Kolodny Jr. Known as a Godley type, they were made in North Central Texas from 1600 B.C. to A.D. 1000.

Modern Native American History

The Wichita were a large group of several hundred thousand Indians of different tribes that shared a common language, living in what is now Kansas and Oklahoma. One band of Wichita migrated south into the area north and south of the Red River in southern Oklahoma and North Central Texas. These were the Southern Wichita, Waco, and Tawakoni groups. The Keechai and Caddoans are also closely related.

The Wichita were excellent hunters and fine gardeners, cultivating large, sometimes fenced gardens of corn, squash, pumpkins, and tobacco. The gardens were usually in the sandy bottoms of creeks and rivers, sometimes several acres in size. They were competent potters, but this activity disappeared after contact with European traders. Their homes were grass lodges, bundles of prairie grass tied to a framework of poles in a beehive shape. The houses were fifteen to thirty feet in diameter, depending on the size of the family. Construction of the lodges was a village effort. Permanent low bed frames attached to the walls and covered with buffalo hides surrounded a fire in the center.

The Wichita life was seasonal. From March through the summer, they were in their village, growing summer vegetables and harvesting mussels and turtles from the creek. By October, the hunters were following the buffalo on the Grand Prairie, west to the Wichita Falls area. At this time, they used conventional tipis and harvested the products

Wichita Indian grass house with grass arbor for food storage. Made of prairie grass over a pole frame, it was the typical summer home of the prairie-dwelling Indians. Photo courtesy of the Western History Collection, Oklahoma Historical Library.

of the hunt all winter. For a thousand years, the Wichita people successfully exploited the Prairie-Cross Timbers junction. This unique combination allowed them to be more successful than tribes who were dependent upon only one source of subsistence.

Indigenous Wildlife

The Native Americans living near the Mound enjoyed an abundant natural bounty to use for food, medicine, clothing, and implements. Among the large game animals were bison, elk, deer, and pronghorn antelope. Large predators included mountain lions, grizzly and black bears, bobcats, wolves, coyotes, fox, and even exotic ocelots and jaguars. Javelina, mink, river otter, ringtail, and badger were also common. Large numbers of turkeys lined the creeks at night. Prairie chickens and grouse were common on the prairie, as well as passenger pigeons. Alligators were abundant in the local rivers and large creeks. The creeks also provided turtles, clams, and mussels. Fish were numerous but not used by the Wichita. The prairie provided natives with as many as 150 different food plants, and 250 plants had medicinal properties. Acorns were plentiful and made into a meal. Other useful trees were hickory, pecan, walnut, persimmon, and wild plum. Osage orange, bois d'arc, was exploited as the wood for bows, and dogwood and cane were used for arrows.

Bison

The Mound lies east of the north-south escarpment that formed the natural barrier for many species, such as the buffalo, and for tribes, such as the Comanche, in the early days. The settlers, as the Wichita before them, made the fall trip to the Wichita Falls area for winter buffalo meat, and in 1862, two local brothers, John and Richard Cristal, were captured by Comanche and never returned. Hall Medlin killed the last buffalo in this area in 1856 on Grapevine Prairie. Mr. Medlin was badly gored but survived the inci-

These bison are part of the herd belonging to John Todd on Wichita Trail. They are descendants of the three hundred survivors of the original sixty million buffalo that roamed the West.

dent. After the Civil War, a bill was presented before the Texas Congress to save the remaining buffalo. General Philip Sheridan, in charge of troops in Texas at the time, testified that the buffalo should be destroyed in order to control the Indians. The bill to save the buffalo was defeated.

European Contact

The first European contact for the Wichita was with the Spanish explorer Coronado in 1541 and Onate in 1601. The first traders to arrive in the North Central Texas area were French from the Mississippi and Louisiana settlements, bringing European trade goods to exchange with the Wichita for furs. In 1719, Bernard de la Harpe was given the rights to trade in the Red River Valley in North Central Texas. In 1758, De Mezieres' expedition passed twenty miles west of the Mound on its way to the Red River. They visited a Keechai, two Tawakoni, and two Taovaya villages in this area. By 1759, the French had established a supply point in Montague County, northwest of the Mound on the Red River. Later, this area became a part of the Louisiana Purchase. Early settlers found these ruins and, assuming they were Spanish, called it Spanish Fort, by which it is known today.

While this contact brought trade goods that eased the lives of the Wichita, it also brought measles and smallpox diseases, which greatly reduced their population. This led ultimately to their near extinction in the nineteenth century when they conflicted with Texas settlers and the harsh policies of the Lamar administration of the Republic. In 1846, after the admission of Texas into the Union, the tribe came under the protection of the United States government. The remainder of the tribe lives in present-day Anadarko, Oklahoma.

First American Settlers

After Thomas Jefferson's Louisiana Purchase in 1803, the American traders began to venture into the region. The first area to be explored was along the Red River, and in 1837, the Holland Coffee Company opened a store at Preston Bend on the Red River. As the Cross Timbers hampered east-west travel, the easiest route to immigrate into North Central Texas was from Oklahoma, south through the low water crossing at Preston Bend.

When settlers first arrived at the Mound, it was under the rule of the Mexican government, in an area called Red River County. In 1837, the Mound became part of Fannin County. The new Congress appropriated funds for 180 Rangers to protect early settlers from Indian attacks. A Ranger station was located at Hickory Creek, south of present-day Denton. In 1844, the Eighth Congress passed an act to establish the Central National Road of the Republic of Texas. It began at the Preston Bend store on the Red River and continued south to the Trinity River, crossing near present-day Dallas and then on to Austin. This became known as the Preston Road and is still known as such today in Dallas.

In 1847, the new state passed an act granting 640 acres of land to married settlers and 320 acres to single men to encourage immigration. W. S. Peters received a contract to manage the sale of much of North Texas known as the Texas Agricultural Commercial and Manufacturing Company of the Peters Colony. The Peters Colony was the first and largest of the empresario form of land grants, including all of Denton County and all or part of twenty-five other counties.

As the American settlers began moving in, the Indians were pushed farther west. What the settlers found was a lush wilderness preserved for them as the Indians had enjoyed it. Wood for cabins and fuel was plentiful in the Cross Timbers. The prairie provided rich grazing for stock and later was plowed for crops. Wild longhorn cattle could be gathered freely. Wild mustang horses were also captured but with much more effort. Until crops were in, settlers lived well on game. One settler reported seeing as many as five hundred deer, antelope, elk, and buffalo grazing within sight of his cabin.

The Prairie and the Settler

Prairie grass was of great importance to the early settlers. The primary occupation of these pioneers was as stockmen. Milk cows and oxen were brought in and mixed with the wild longhorns. The prairie was considered free grazing for all. The prairie was extremely abundant and productive, creating an incredible fire hazard in dry weather. Lightning started accidental fires, and the Indians often set fires intentionally to improve the grasses and attract game to the new growth. A man on horseback could not outrun a full-blown wildfire, and this was one of the greatest dangers to settlement. In 1846, a wildfire that began near present-day Dallas was fueled by southerly winds and burned everything to the Red River. One of the first laws passed by the Texas Congress in 1848 controlled the season and methods by which prairies could be burned. Many farmers plowed firebreaks around their houses as a precaution. As dangerous as it was to settlers, the regular burning of the prairie was nature's way of regulating the health of the plants and destroying competitive woody invaders. Interruption and suppression of the burn cycle was the first damage done by settlers to the Great Prairie.

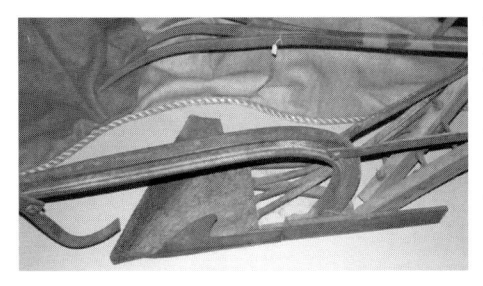

This plow was invented to remove the top layer of grass from the prairies, enabling ordinary plows to turn over the virgin land underneath. The sod created was used to build temporary homes on the treeless prairie. Photo was taken at the Sternberg Museum, Hays, Kansas.

The next phase of the extermination of the Great American Prairie began with the invention of the steel prairie plow. Before, the plow boards could not turn over the dense root systems of the grasslands. When it was discovered how very rich these grasslands were, it was just a matter of time until almost all lands were under cultivation. However, a few European farmers understood the wisdom of retaining certain tracts of native prairie for hay in hard times. Prairie grass was highly productive, did not have to be sown or plowed, and was dependable. This is the manner in which the Mound and a few other acres were preserved.

B

HATTIE O.	GEORGE L.
WIFE OF	BEAVERS
G. L. BEAVERS	BORN
BORN	FEB. 15, 1829
FEB. 9, 1835	DIED
DIED	OCT. 5, 1906
SEPT. 21, 1910	

God is our refuge and strenght
a very present help in trouble.

BEAVERS

Chapter 2
Ownership of *the Mound*

John and Edy Wiswell

One of the first to take advantage of the free 640 acres of land offered by the state was John R. Wiswell and family. In 1844, John, his wife, Edy, and their five children arrived from Illinois to fulfill their part of the bargain. They had to build a home, fence and cultivate fifteen acres, and live on the tract for three years in order to claim ownership. On the south edge of their property was a curious rise in the prairie, soon to be called the Flower Mound. John Wiswell died only two years after arrival, and Edy married a local doctor, Burnett J. Doen. When Dr. Doen passed away around 1854, Edy again remarried to a Mr. Higgins. No gravesites or descendants of the Wiswells have been located in Denton County to date.

George and Hattie Beavers

On December 23, 1857, Edy Wiswell and two of her children, John and Henry, sold their shares of the tract, along with the shares of a deceased son, Isaac, to George L. Beavers. The Mound was a part of this sale.

George L. Beavers was born on February 15, 1829, in Sweetwater Valley, Monroe County, Tennessee. On March 5, 1857, George married Harriet Owen, also of Monroe County. Harriet, called Hattie, was born February 9, 1835, to Charles and Louisa Berry Owen. George and Hattie welcomed their first child, Spencer O. Beavers, on their new Texas farm on April 11, 1858. Spencer passed away on October 23, 1881. Other children were William Hartsell Beavers, 1859–1936; Fronia Beavers; George H. Beavers; Charlie Beavers; and Ruth Beavers.

George and Hattie Beavers purchased Joseph and Sarah Wiswell's tract on August 3, 1872. On July 30, 1873, James Bingham and Elizabeth Bingham, heirs of John Wiswell, also sold their shares to Mr. Beavers. Henderson Murphy, another heir, sold one tract to Mr. Beavers on August 27, 1873, and his last tract to Mr. Beavers on June 21, 1875. On February 10, 1876, George Beavers bought the share of V. S. Vanhoose and his wife, J. D. At last, George and Hattie Beavers had accomplished their goal of reassembling almost the entire John R. Wiswell Survey. Daughter Ruth married R. L. Donald and, along with W. H. Cowan and his wife, H. S., and W. H. Beavers, inherited the Mound from George and Hattie. On June 21, 1876, all this was conveyed to C. F. Beavers and Belle Beavers. When C. F. passed away on August 11, 1910, Belle conveyed the land to their daughter, Charlie Fay Beavers.

The gravestones of George and Hattie Beavers and Belle and Charlie Beavers are in the Beavers' family plot in the cemetery of the Flower Mound Presbyterian Church.

George Beavers was a prominent member of the early rural community, being one of the principal founders of the Flower Mound Presbyterian Church. In 1877, he was one of the four founders of the Donald Academy, the first school in Flower Mound, built just north of the Mound on what is now FM 2499. He is believed to be the George L. Beavers listed as a private in the Confederate Army,

Thirty-fourth Texas Cavalry. George L. Beavers passed away on October 5, 1905. Grandson Earl Beavers and great-grandson Gary Beavers still survive him in the Flower Mound area.

Ray and Charlie Fay Lester

Ray B. Lester was born at Chinn's Chapel, January 15, 1898. In 1906, the Lesters moved to Lewisville. Later, Ray went into business with his father, "Bent" Lester, as T. B. and Son Grocery and Market on Main Street in Lewisville. The store operated there for forty-one years. In 1917, Ray married Charlie Fay Beavers, and they moved onto the old Wiswell farm that Charlie Fay had inherited. In 1936, they built a rock house on Main Street in Lewisville and moved to town. They continued to commute to the land, fencing the entire tract and raising cattle. They owned the Mound for forty-two years, during which time, like all of the owners in the past, the native prairie was saved as hay meadow and never plowed. They had one daughter, Ruth Rozell.

Ray and Charlie Fay Lester rest during a hunt on their property, which included the Mound in the early 1900s. Photo courtesy of Sweet Flower Mound Land.

Edward and Betty Marcus
Black Mark Farms

In the early 1950s, Edward and Betty Marcus began investing in land in Denton County. Edward was the chairman of the board for Neiman-Marcus, the famous clothing store in Dallas, with his brother, Stanley Marcus. Betty Marcus was of the Blum Candy family in New York.

Edward became a gentleman rancher who loved the land and raising cattle. He called his venture Black Mark Farms. The farm grew to include almost four thousand acres of land in southern Denton County. The man he chose to oversee this venture was Bob Rheudasil. In Bob's competent hands, they increased the herd of champion Angus cattle to over one thousand head, improved the quality and production of the land, and planted thousands of trees on their tree farm. They imported a registered bull from New Zealand to increase the quality of their herd and won many awards for prize heifers. They kept the Mound as a hay meadow. It was Bob Rheudasil who negotiated the sale between Ray Lester and Edward Marcus on February 16, 1959. After this time, the cattle were sold and the direction changed to buying land for future development.

Edward and Betty Marcus in western attire on the top of the Mound in the 1950s. The camera is facing east toward Black Mark Farms. Photo courtesy of the New York Times.

Stanley Marcus on Edward Marcus

"My brother, Edward, was never quite sure that he wanted to be a merchant, but he was quite certain that he wanted to be a farmer, rancher, and town planner. Actually, he was a gifted merchant, with a fine sense of taste, a keen appreciation of quality, and a superb talent in relating to the people who were his customers and employees. He got his kicks, though, when he put on his Levis and his

Chapter 2 Ownership of the Mound

Part of the Marcus' large herd of one thousand Angus cattle grazes the Mound in the 1950s. Photo courtesy of Black Mark Farms.

Award-winning heifer. Center, Edward Marcus, next to Betty Marcus, and far right, Bob Rheudasil, property manager. Photo courtesy of Black Mark Farms.

Indian Paintbrush flourish on the Mound around 1970. Bob Rheudasil seeded these. Photo courtesy of Black Mark Farms.

Aerial view of the Mound in 1971, including the intersection of FM 2499 and FM 3040. Photo courtesy of Black Mark Farms.

boots to go out to his Black Mark Farms on a Friday afternoon for the weekend."

Flower Mound New Town

In 1970, the Congress of the United States enacted Title VII of the Housing and Urban Development Act, intended to provide low cost loans to encourage innovative solutions to the problems of planned urban communities. Mr. Marcus liked the idea and began to pursue assembling the land and organization required to participate. Almost six thousand acres were assembled to be dedicated as Flower Mound New Town, with Black Mark Farms as the core. HUD provided $18 million in guaranteed loans in 1971, and Flower Mound New Town became the fourth project to receive the New Town designation. Raymond Nasher, developer of North Park in Dallas, was brought in as managing director. The project was complicated and expensive and, like most of the New Town projects, failed when the Nixon Administration withdrew support of the concept and HUD foreclosed on the notes. The First National Bank of Dallas then became the owner of the Mound on September 7, 1976.

Bellamah Community Development

The First National Bank of Dallas and HUD continued to keep the project going under the management of Marty Matyas. The trees Mr. Marcus and Mr. Rheudasil had planted were now grown and provided a source of income to the bank during the transition. Ultimately in 1981, HUD, at Matyas' suggestion, placed the remaining three thousand acres for sale to the highest bidder over a minimum bid of $15 million. Bellamah Community Development of Albuquerque, New Mexico, was the only bidder at that price. The deed was filed in Denton on February 4, 1982. Mr. Matyas stated that the Bellamah people arrived with two bids, one for $15 million and one for $20 million. They waited until the last minute, and since no other competitive bids were made, they were secure in handing in the minimum bid. If the bank had told them to turn in their bid and leave, it would have secured an extra $5 million for the government.

2004 aerial photo of the Mound and 2499-3040 intersection showing urbanization over the course of thirty years. Photo copyright 2004, Town of Flower Mound and the North Central Texas Council of Governments.

Chapter 2 Ownership of the Mound

Looking east from the Mound in 1970. Flower Mound Farms is in the background.

The same view in 2003, showing increased population density over the course of thirty years.

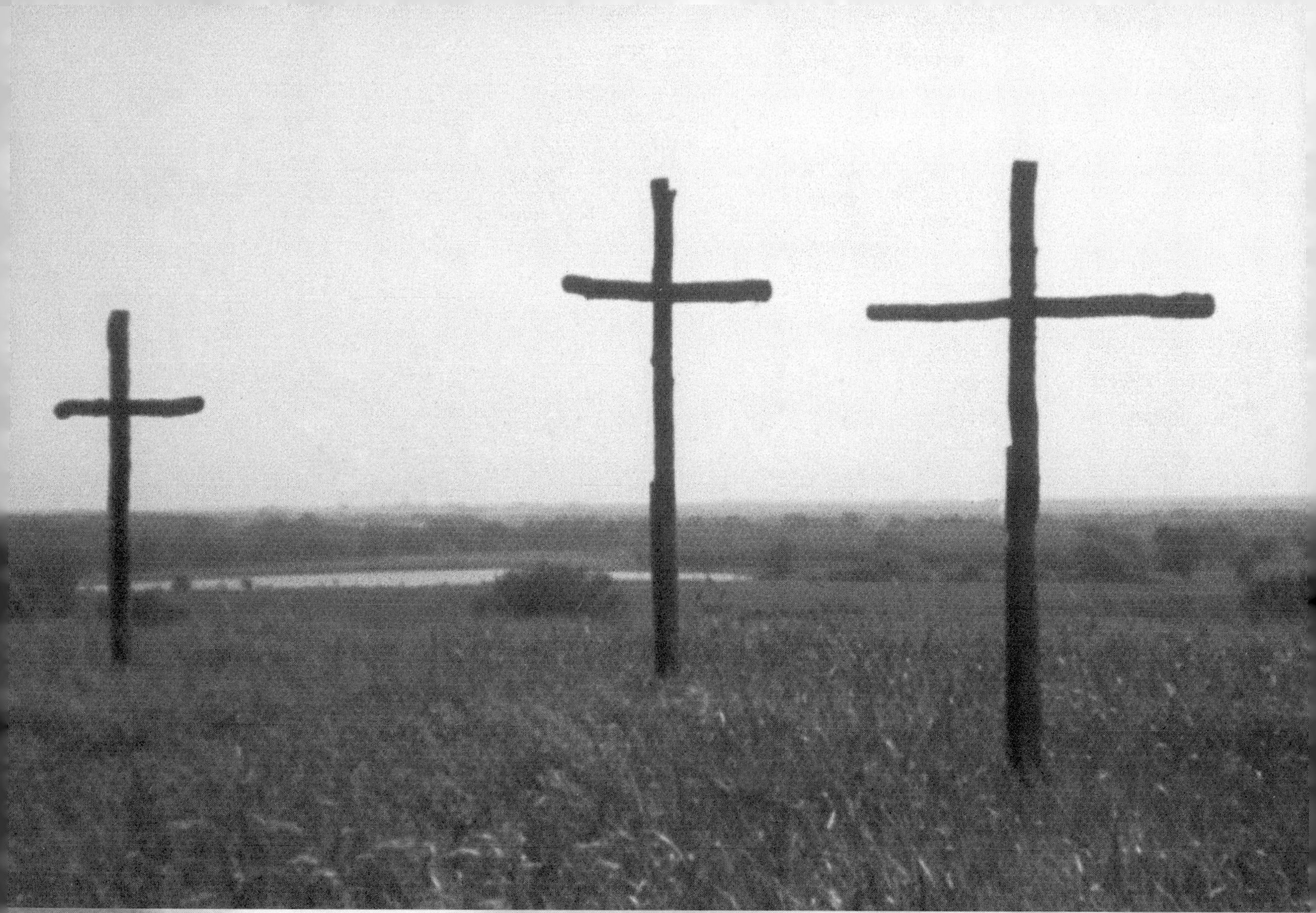

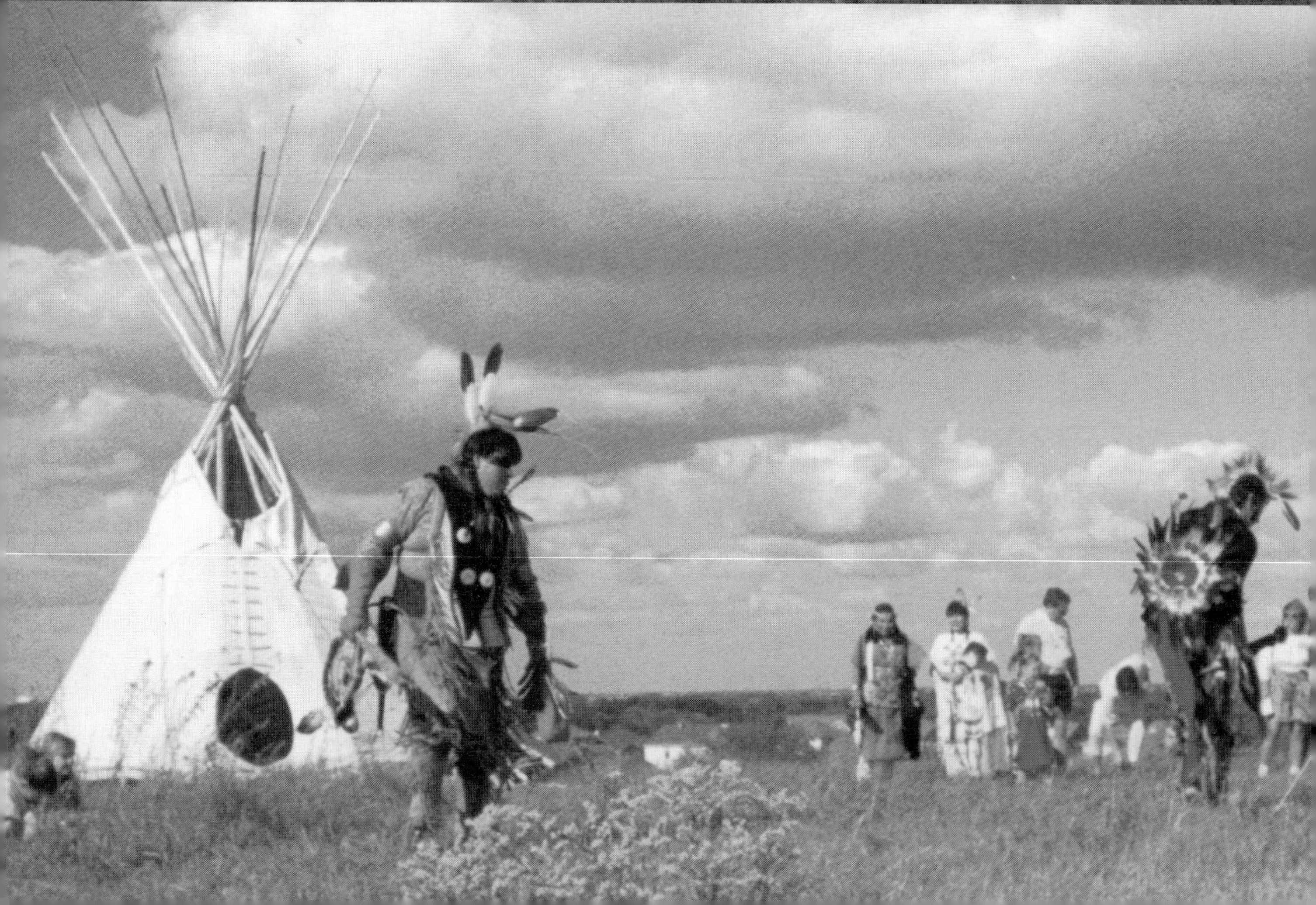

Opposite Page Top: The first Easter sunrise service in modern times was held on the Mound in 1971. The Summit Club and the Women of Flower Mound sponsored it. Notice the unmowed prairie grass. Photo courtesy of Black Mark Farms.

Opposite Page Bottom: Wichita Indians from Anadarko, Oklahoma, reconsecrate the Mound and provide entertainment at the first Prairie Picnic in October 1988. Photo courtesy of JoAnn Hanson.

Chapter 3
Related History

Religious Services on the Mound

The first recorded religious service in Denton County was a sermon preached by John B. Denton on Long Prairie in May 1859. Since the Mound is a part of Long Prairie, it could have happened on the Mound, as it is a natural site for such activity. After this time, the Mound was frequently used for camp meetings. Since there were no churches yet, the early settlers would hire a traveling minister, and all the farmers in the area would meet at a natural landmark, like a grove or creek, and camp for a week or more for spiritual communion. At the same time, it was a chance for bartering, visiting, and selling crops and animals, and since the closest settlements were Lewisville and Grapevine, the Mound was a central meeting place in this area.

In 1971, the Summit Club, a Flower Mound men's club, and the Women of Flower Mound Club sponsored the first nondenominational Christian service on the Mound in many years. It was an Easter sunrise service, complete with three old rugged crosses. The tradition has continued to the present; over a thousand people attended the 2003 service.

The Mound is open to all religions and groups upon request and has been the site of Jewish services, religious observances of the Bahai faith, and Native American consecration services. There have been several weddings on the Mound and at least two memorial services.

Good Neighbors

The Donald Community

In 1877, when George and Hattie Beavers owned the Mound, the Donald family owned the land to the west and south, an area known as the Donald Community. That year, Mr. Beavers, R. H. Donald Sr., and Jess Crawford contributed $192.50 to build a schoolhouse to replace the old log schoolhouse east of the Mound. J. H. Donald donated the land just west and a mile north of the Mound, and they erected a two-story building called the Donald Academy. The new school was the largest rural school in Denton County and operated into the 1940s when the schools came under the jurisdiction of Lewisville School System. One hundred years later, in 1986, the Lewisville Independent School District honored the Donald family by naming the new elementary school just north and east of the Mound Donald Elementary. The last direct descendant of the original Donald family was R. L. Donald Jr., who passed away in Lewisville in 1995. He was the son of George and Hattie Beavers' daughter Ruth and her husband, R. L. Donald Sr.

The second Donald School, called the Donald Academy, was constructed near the Mound in 1877 for $192.50. This photo was taken in 1933. Photo courtesy of the Lewisville Independent School District.

25

The Flower Mound Presbyterian Church

In the first half of the nineteenth century, the area including and a few miles surrounding the Mound was referred to as Long Prairie. This name survives as the local name for FM 2499, Long Prairie Road. After 1850, the name Flower Mound became more popular. The first recorded use of the name Flower Mound was with the forming of the Flower Mound Presbyterian Church in 1854. Mathew Lyle Donald was the first minister. The group met in homes at first until the log church could be built. Later, in 1879, Mr. Donald officially donated the ten acres of land containing the log church, permanent brush arbor, and cemetery. The brush arbor is one of the few remaining structures of its kind in Texas and is slated for restoration. The cemetery is one of the most important historic sites in the area and is the final resting place for many previous owners of the Mound and its neighbors. The Donald and Beavers families are there, as well as Otto and Babe Consolvo. There are 450 markers in the cemetery, but it is estimated to contain as many as two thousand graves. A thorough history of the church is included in the A. C. Greene report on Flower Mound, available at the Flower Mound Library.

The Alexanders

The Alexander family occupied the house closest to the Mound, across FM 3040 and to the east a few hundred yards, until the early 1980s when it was torn down.

Voluntine Alexander brought his family to the Flower Mound area in 1884 from Alabama. His son, James Decatur Alexander, was ten years old at the time. Voluntine died on February 28, 1897, and is buried in the cemetery of the nearby Presbyterian Church.

On June 16, 1889, James married Lamanda Florinda Clementine Hurley, called Clem. Jim and Clem had twelve children from 1900 to 1925: a baby girl who died at birth, and Ruby, Alberta, James Russel, Jewel, Ned, Fannie, Cleo, Ollie, Clemey, Ray, and Milton. Four of these survive at this publication: Ned, 95; Clemey, 86; Ray, 83; and Milton, 79.

Top: The Flower Mound Presbyterian Church, established in 1854, Mathew L. Donald, Minister. This is the first formal use of the name Flower Mound. The structure dates to 1901.

Bottom: The permanent brush arbor at the Flower Mound Presbyterian Church is a rare remnant of early Texas religious services. Before churches were built, temporary brush arbors were erected to protect the members from the harsh Texas sun at camp meetings. More permanent arbors were created later and are similar to local Indian prairie grass arbors (see the photo of the Wichita Indian grass house). Church members refer to it as the "Tabernacle."

Several children and grandchildren were born in this house, including Florence Clementine Alexander, oldest child of James Russel (Jimbo) and his wife, Leona Feemster Alexander. Jim plowed his fields behind mules and planted cotton, peanuts, and row crops. Clem had milk cows, chickens, and a vegetable garden. Clem planted her own cotton patch for a fine grade of cotton for quilt batting, as she was an excellent quilter. She also made most of the family's clothes.

The couple lived on this farm until their passing—Clem on December 10, 1958, and James on March 30, 1967. The farm was divided among their children and later sold for development.

The wedding pictures of James and Clementine Alexander, taken on June 16, 1899. The Alexanders lived in the house closest to the Mound, across FM 3040, to the south and east. Photo courtesy of Bob and Florence McCarley.

26

Florence married Bob McCarley, and they reside in Paris, Texas. Bob and Florence McCarley were generous contributors to the fence completion project and this book. A bronze plaque is mounted near the east gate with an image of the Alexander home.

Mound Legends and Lore
Tall Tales

Many stories have been passed by word of mouth about the Indians and the Mound. Basically, they were warnings to the settlers that the Mound was a sacred burial site and nothing should ever be built on the Mound or violators would suffer dire retribution. These stories have some basis in fact. For instance, there is evidence that the Wichita preferred to bury their dead in shallow graves on hilltops overlooking their villages. Since the Mound is the highest natural site in close proximity to known Indian camps on Denton Creek, it is likely that the Mound does contain ancient burials.

Ben Lester is said to have brought materials to the Mound to build a house, and they were strewn about by a tornado. He moved everything to the north end of the Mound and built his house there. This house was removed later. The Flower Mound Presbyterian Church is also said to have tried to build on the Mound with similar results, and legend says they then moved to the present location on FM 3040. As previously stated, this was in 1854 and documents the first formal use of the name Flower Mound. Otto Consolvo used these stories to great effect in his campaign to preserve the Mound in the 1980s.

There are accounts that, after the Civil War, Comanche raiders from reservations in Oklahoma came into the area near the Mound. They would raid during the full moon, hiding out during the day in the Cross Timbers and stealing horses at night to drive back to Oklahoma. Once, they stole a fourteen-year-old boy from a farm on Long Prairie just east of the Mound. He was never seen again. In another occurrence, a Mr. Fortenberry followed the raiders too closely, and they turned and killed and scalped him. These events kept the settlers wary until regular patrols by the Rangers provided stability in the late 1800s.

The intersection of FM 2499 and FM 3040 has always been a central location in the local farming community for rendezvous. "I'll meet you at the crossroads" was the expression often used. The shopping center on the northeast corner next to the Mound is called Flower Mound Crossing. It is said that when young people wanted to elope, they would meet the preacher at the crossroads, get married, and then go home and tell their folks. Nearby neighbor Mr. Simmons had another story of a meeting at the crossroads in the twenties. The roads were narrow dirt affairs at this time, and he was waiting for a friend late one afternoon when he fell asleep in the middle of the crossroads. He was awakened at dusk by a curious possum in time to see the approaching headlights of his friend's Model A.

FLOWER MOUND

SETTLERS OF THE PETERS COLONY NAMED THIS SMOOTH, DOME-SHAPED HILL FOR THE ABUNDANT WILD FLOWERS THAT GROW ON IT. RISING FIFTY FEET ABOVE THE SURROUNDING PRAIRIE, FLOWER MOUND LONG HAS BEEN A POINT OF INTEREST IN THE AREA. ACCORDING TO LOCAL LEGENDS, NO STRUCTURE WAS EVER CONSTRUCTED ON TOP OF THE MOUND, NOR HAS ANY TREE GROWN HERE.

BEFORE W. S. PETERS BEGAN BRINGING SETTLERS TO THE LAND ISSUED HIM BY THE REPUBLIC OF TEXAS CONGRESS, WICHITA INDIANS INHABITED THE AREA. DURING THE 1840s, PETERS COLONISTS BEGAN MOVING TO THE PRAIRIE IN SEARCH OF GOOD FARMLAND. IN 1844, JOHN R. WIZWELL WAS GRANTED 640 ACRES OF LAND THAT INCLUDED THE MOUND. HIS WIDOW, EDY, LATER REMARRIED AND SOLD THIS LAND TO GEORGE L. BEAVERS. FLOWER MOUND REMAINED IN THE BEAVERS FAMILY WELL INTO THE TWENTIETH CENTURY.

ALTHOUGH THE HILL HAS REMAINED IN PRIVATE OWNERSHIP, IT HISTORICALLY HAS BEEN IDENTIFIED WITH THE COMMUNITY THAT GREW UP AROUND IT. FLOWER MOUND PRESBYTERIAN CHURCH WAS THE FIRST TO OFFICIALLY USE THE NAME IN 1854. ONCE A SPRAWLING AGRICULTURAL COMMUNITY, FLOWER MOUND HAS BEGUN TO EXPAND WITH THE URBAN GROWTH OF NEARBY DALLAS AND FORT WORTH, LEAVING THIS FORMATION AS A HISTORIC REMINDER OF ITS PIONEER DAYS.

(1984)

Chapter 4
The Mound Corporation

Otto Consolvo, Founding Chairman

The last step in the saga of permanent protection for the Mound began with Otto Consolvo. Born June 6, 1900, Otto Consolvo was a colorful man who was civic-minded and devoted a lot of time to the newly created Town of Flower Mound. Otto was passionate about the Mound. His wife, Babe, was a full-blooded Cherokee Indian from Oklahoma, and it was through this heritage that she was attracted to the Mound. Otto and Babe would ride horses to the Mound and find arrowheads and pottery shards. It became Otto's personal mission to protect the Mound from destruction, especially after Babe passed away in 1982. Doggedly pursuing this goal, he finally was able to convince Bellamah Corporation to sell the 12.576-acre site to the citizens of Flower Mound for ten dollars on September 2, 1982. Bellamah required that, first, in order to increase protection for the Mound, the property be given to the citizens of Flower Mound and not the town government, and, second, the perimeter was to be fenced to control access. A board was formed to manage the property in perpetuity. Otto organized the Mound Corporation and served as the first chairman. In addition to Consolvo, other original officers were Edith Bettinger, secretary; Ben Campbel, treasurer; and trustees Gregory S. Carter, Ronnie Hilliard, Mary Webb, Arlene Wise, JoAnn Webb, Willard French, and Ernest Hilliard, honorary lifetime trustee.

Otto and Babe Consolvo. Without their intervention, the Mound would probably have been leveled and covered with homes.

The founding board of the Mound Corporation. Chairman Otto Consolvo, center left, receives a $7,500 check from Suburban American representative James Brickman, center right, to help construct the perimeter fence. From left are Ben Campbel, Arlene Wise, Edith Bettinger, Otto Consolvo, Ronnie Hilliard, James Brickman, Gregory Carter, Ernest Hilliard, and JoAnn Webb. Photo courtesy of JoAnn Webb and Tommy Webb, mayor of Flower Mound, 1970–1976.

The Mound Corporation bylaws were drawn up, and Jack Wise and Jerry John Crawford handled legal affairs. Immediately, and for quite some time later, their attention was devoted to securing a tax-exempt status for the Mound with the Internal Revenue Service.

The First Fence

Otto was occupied with raising the money needed to fence the perimeter of the Mound property. Suburban American Corporation, the company managing Lake Forest Development, of which the Mound would have been a part, donated $7,500 to this campaign. A three-rail cedar fence was installed completely around the twelve-acre site.

29

The Historic Marker

A five-hundred-word document about the Mound was prepared with the Denton County Historical Commission and sent to the Texas State Historical Commission for approval. The 27x42-inch bronze plaque was cast in San Antonio and arrived in 1984.

According to the plaque, no trees have ever grown on the Mound. However, this refers to the old days. Bob Rheudasil planted the live oak at the front and the Arizona ash trees on the east side for Edward Marcus in the 1950s.

The Flower Mound Monument was constructed in 1985 by Henry "Chief" Benson to showcase the Texas historic marker denoting the Flower Mound as a historic site.

The Monument

A fundraiser was held to provide materials for a proper stone monument to display the new state historic marker. Henry "Chief" Benson, a Lewisville stone mason, volunteered to build the structure. Eight feet tall and eighteen feet long, the monument was expertly fabricated of cut sandstone. A plaque on the reverse side lists the names of the donors. The monument was completed in 1985.

Chief Benson passed away on January 1, 1999. Born in 1917 of Cherokee Indian heritage, he and his wife, Rita, had three daughters and one son. He learned his trade in the CCC (Civilian Conservation Corps) and later worked on the chapel at Texas Woman's University in Denton. He also created the original Veterans Memorial in Lewisville, the flag monument at Lewisville City Hall on Main Street, and the sign at Flower Mound Cemetery. His favorite works were the sign at the Lewisville Bible Church where he and his family attended and, of course, the Flower Mound Historical Monument. Chief wrapped pictures of his children and secured them in the monument.

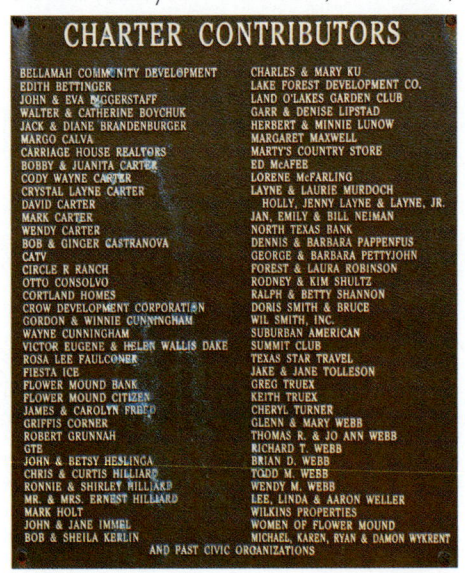

Portrait of Henry "Chief" Benson. Mr. Benson created the Flower Mound Historic Marker Monument in 1985. Photo courtesy of Mrs. Henry Benson and daughter Nelda Ebsen.

Ronnie Hilliard Chairmanship

The Wrought Iron Fence

After a few years, it became apparent that the cedar rail fence was inadequate protection and would require a great deal of maintenance. The Trammel Crow Company was interested in securing a commercial zoning ordinance for the property west of the Mound, now Flower Mound Crossing shopping center. In 1986, the Trammel Crow Company donated three sides of the wrought iron fence that stands today. The frontage cedar rails were left in place. The cedar rails that were removed from the other three sides were sold to raise funds.

The bronze plaque on the reverse of the monument honors the donors of the first fundraiser who provided an operating fund for the Mound Corporation.

Bill Neiman Chairmanship

John Immel, Rosa Lee Faulconer, Doris Smith, Ronnie Hilliard, Betsy Heslinga, Forest Robinson, Ralph Shannon, Barbara Pettyjohn

When Ronnie Hilliard resigned as chairman, board member Bill Neiman was elected. Bill and his wife, Jan, owned Neiman Environments, a nursery and landscaping company located in Flower Mound that focused on drought-tolerant native plants. Bill harvested the first native seeds from the Mound that were shared with other prairie restoration projects. A trained and registered prairie burn specialist, Bill consulted with federal and state range experts on the condition of the Mound, which resulted in a recommendation to perform a controlled burn.

Mound Chairman and Fire Boss Bill Neiman sets fire to the Mound during the 1988 controlled burn. Photo courtesy of Native American Seed (www.seedsource.com).

Considered the best choice for removing thatch buildup and reducing woody invasion, the prairie was managed by fire for thousands of years, by natural and human means. The Mound was burned with permit and without a problem. In 1986, the Lake Forest Development was completed with homes adjacent to the east and north sides of the Mound, and when time came for another burn, the neighbors raised concerns. A public hearing was called, and much evidence was presented to the City Council regarding safety and the need for the burn. Most fears were allayed, and the council voted to grant the permit. One of the neighbors even became a trustee. The burn was executed successfully again, with the help of the Argyle and Flower Mound Fire Departments. Neighbors claim the flower show on the Mound the next spring was the best they had seen. Attempts to continue burning the Mound have met strong opposition in recent years. City Council has refused to grant burn permits because of increased population density and insurance issues. The woody plant invasion increases with each year the Mound is not burned, and it is hoped someday a solution can be found to resume periodic burns. There is a five-acre prairie in a Chicago suburb that is safely burned every year.

The Argyle and Flower Mound Fire Departments backed up volunteers manning the firebreaks. Fire-retardant chemicals were spread on the perimeter to successfully contain the blaze. Photos courtesy of Native American Seed (www.seedsource.com).

The Prairie Picnics

Under Bill Neiman's chairmanship, two large events were undertaken to promote the Mound and prairie awareness. They were called the Prairie Picnics and occurred on the first Saturday in October in 1988 and 1989. The picnics were well received and great family fun.

Indians from the Wichita tribe remaining in Anadarko, Oklahoma, attended and danced in costume. Spectators joined the dancers. At the second picnic, a raptor demonstration was added. John Swenson was added to the board and was elected chairman in 1991.

Wichita Indians from Anadarko, Oklahoma, provide entertainment at the first Prairie Picnic in October 1988. Photo courtesy of Doris Smith.

Alton Bowman Chairmanship

John Swenson, Lori Wilkins, Dave Wilkins, Anita Leone, John Nelson, Joan Nelson, Andre Gerault, Levenia Gerault, Andrew Eads, Bruce Thompson, John Sullivan, Nicole Dogan, Nelson Ringmacher, Albert Picardi, Jack Wise

Fence Completion

The frontage of the Mound was left in cedar rail fencing. After ten years of use, this declining cedar rail fence needed replacement, and another fundraiser provided the twelve thousand dollars necessary to complete the wrought iron fence and gates that stand at the frontage today. The names of those donors are on a brass plaque welded to the small gate near the monument. In 2001, trustees created the signage in large white letters on the front fence publicly identifying "The Flower Mound."

The bronze plaque listing the donors of the fence completion project in 1998 is located at the east entrance to the Mound.

Documentation and Photography of the Plants

With the IRS 501(c)(3), i.e., tax-exempt, status established and the Mound entirely fenced, the last barriers were cleared to implement the other directives of the charter, which are scientific, historical, and educational. In 1998, Alton Bowman began photographic documentation with 35mm slides that spring season. Weekly trips to the Mound were made to capture emerging blooms. Dates were noted, and research was conducted to positively identify each species. This book represents five years of documentation. A full collection of one hundred 8x10 prints was framed, and many prints were sold as fundraisers. It was decided that a history of the Mound would be added to the photos and botanical descriptions of the field guide. One hundred percent of the proceeds of this guide will be used for the perpetual care of the Mound.

Sale of the Frontage for 3040 Expansion

In 2001, the Town of Flower Mound and the Texas Department of Transportation began acquiring increased rights-of-way from properties with frontage on FM 3040 between FM 2499 and Lewisville for road improvements. The Mound had an old survey in which the property line went to fifty feet from the center of the road. This meant that the town and TXDOT must purchase from the Mound Foundation a parcel fifty feet deep by six hundred feet of frontage. Trustee Andre Gerault, senior member of the negotiation team for the Mound, rejected an initial offer of one dollar per square foot. On August 22, 2001, parties agreed on a price of $30,900, and 16,495 square feet of land was conveyed to the Town of Flower Mound. Construction for the new road began in 2003.

Pottery shards found on the Mound by Sweety Bowman authenticated the tract as a Native American archaeological site with an official listing. The shards are from the Archaic period, before A.D. 700.

Archaeology Update

During construction, archaeologists Paul and Jan Lorrain were summoned to examine the possibility of acquiring an archaeological designation for the Mound site. Two pottery shards found on the Mound by Sweety Bowman in the 1970s are the only extant archaeological remnants known from the Mound. They were identified as Archaic period specimens and offered proof that Indians used the

Mound prior to A.D. 700. With this evidence, on March 11, 2004, the Texas Archaeological Research Laboratory at the University of Texas at Austin officially designated the Flower Mound 1 Site as 41DN528. The "41" indicates Texas, "DN" signifies Denton County, and "528" refers to the 528th site registered in Denton County. The location of the nucleus of the site is 680551E/3654400N.

Gilgai (Hog Wallows)

Unique to black land prairies is a phenomenon called gilgai, also descriptively known to settlers as hog wallows. They are present on the Mound and are one of the identifying characteristics of native prairies. Gilgai are shallow depressions caused by repeated drying and wetting of the clays. Large cracks, up to six inches wide and at least a foot deep, created in the dry season, made horse travel at night dangerous for the settler. The cracks filled with loose dirt, and then in the rainy season, this material would swell and push out the sides, creating a depression. On slopes, they create a washboard effect. Gilgai are evident on the west side, immediately after entering the small west gate. They continue north to the northwest slope of the Mound where small ridges are formed. These basins create a diversity of plants as they catch and hold water for some time. All of the Wild Hyacinth grows in these hollows, as well as some of the Prairie Rose.

Remaining Wildlife on the Mound

The Mound contains only a remnant of the vibrant diversity it once had as a part of the great prairie ecosystem. Now an isolated fragment, it is questionable if it can even survive without that connection. However, it is still much more diverse than anything around it. Coyote and raccoon scat is not as numerous as even a few years ago but is still observed. Owl pellets are sometimes left on the monument, and red tail hawks and marsh hawks, harriers, are regular visitors for the numerous rodents, especially after the mowing. The high yield of seeds is responsible for the large number of prairie rats and field mice. Not surprisingly, one copperhead and one rat snake have been listed. The insect population is high, with lots of pollinators, bees, wasps, and butterflies. Many insects, like the yellow swallowtail butterfly, are associated with prairie plants, and the yucca is only pollinated by the yucca moth. Scissor-tailed flycatchers in courtship flight are common in the spring. In the fall, waves of mourning dove come for the sunflower seeds. One box tortoise has been listed, and there are active burrows occupied by nine-banded armadillos. Regardless of the decline, the Mound is still an inspiring place to enjoy nature in an urban environment.

Some of the diverse life remaining on the Mound: the three-toed box turtle, Terrapene carolina triunguis, *and the eastern black swallowtail butterfly,* Papilio polyxenes asterius.

Field Guide

How to Use the Field Guide

The plants are arranged in approximate order of flowering during the year. Only plants found to bloom on the Mound are included in this guide. If you want to identify a plant you find on the Mound, look in the chapter that includes the date of your visit. Some plants bloom all summer, so you may have to search more than one chapter. The flowers are also grouped according to color at the end of each chapter. Alternately, if you know the name of a plant, use the index to find its page number, and the text will tell you where and when it blooms on the Mound. Notice the purple variations are listed under Blue.

Engelmann Daisy

Musk Thistle

Big Bluestem

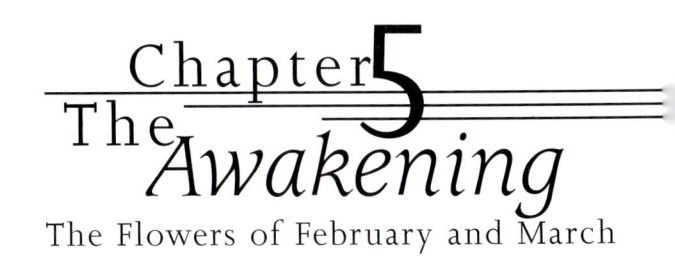

Chapter 5
The Awakening
The Flowers of February and March

This field guide section is laid out in order of bloom time over an entire year, beginning with the earliest bloomers in February to the last show of color in November, before the first freeze. This is the annual life cycle of the Mound. With some adjustment for rainfall, virtually the same thing happens at the same time every year.

Walking the Mound in February is the easiest time of the year. The entire acreage will probably have been mowed closely over the winter, after the first freeze. The dead grass is already decomposing. Days are getting warmer. New bright green shoots are easy to spot against the dull brown of last year's grass. Sometimes, you may have to clear old growth back to see the emerging plants, but they are there. Every warm day brings more color than the day before. The Mound is awakening.

Fringed Puccoon

Ten-Petal Anemone
Anemone berlandieri

FEB–APR

Buttercup family (Ranunculaceae):
Windflower

Description: Our plant grows to 10 inches in height. It is hairy and unbranched. The flowers grow up to 2 inches across. They are solitary on the tip of a long stem and are usually white, although some are pale blue-lavender. The outside of the 10–20 petal-like sepals is often streaked with light purple. The underside of the bottom set of sepals is usually reddish-wine colored. The deeply cut leaves are mainly found at the base of the plant where they tend to be purplish on the underside. There is a whorl of three deeply lobed leaves halfway up the stem.

Location: When in full season, this flower can be found all over the entire 12 acres of the Mound. The best specimens seem to be on the north central section, about 20 yards from the top.

Season: This is the Mound's earliest bloomer, appearing by the first week in March.

Comments: The plant opens late in the day and closes early. It often closes on cloudy days. Anemone is a Greek word meaning "of the wind," hence "Windflower." It gets this name from its seed dispersal system, similar to the dandelion. A month after blooming, the tall fruiting seed heads release tiny singular seeds, floating from delicate parachutes.

Bladderpod
Lesquerella gordonii

FEB–JUN

Mustard family (Crucifer or Brassicaceae):
Popweed; Gordon's Bladderpod

Description: This is a low, sprawling, hairy plant growing up to 12 inches tall. Each plant has multiple stems growing from a common base, forming tufts or clumps. It has bright yellow flowers, growing to 1 inch across, with 4 petals and 6 stamens. They are loosely arranged on an elongated, terminal spike. Its narrow and thin leaves are up to 4 inches long, alternate, and stalked. They grow mainly from the base of the plant. The fruits are pea-sized round pods about 1/8 inch in diameter, on curved stalks.

Location: This plant can be found in numerous locations, especially on the east side, midway along the fence and on the southwest slope of the Mound.

Season: This plant is an early bloomer, emerging almost simultaneously with False Garlic and Anemone. It may rebloom later in the season, after summer rains.

Comments: The Bladderpod flower is prominent, in large clusters. The name comes from the popping sound heard when the fruits are stepped on. There are 15 such Bladderpod species found in Texas. When rains are good, it forms large colonies on the front of the Mound.

False Garlic
Northoscordum bivalve

FEB–MAY; SEP–OCT

Lily family (Liliaceae):
Crow Poison

Description: False Garlic has small white flowers that grow in clusters of 6–12 at the top of a leafless stem that grows to 10 inches in height. Each flower has 6 white, petal-like segments, pointed at the end, with narrow but prominent stripes of green, red, or purple on the outer surface. The flowers grow to 3/4 inch wide and have 6 yellow stamens and anthers at the center. The leaves grow to 16 inches in length and are quite slender and flat. They are found only at the base of the plant.

Location: These plants are easily found just inside the south fence, most abundantly near the west entrance and on the east side of the south slope.

Season: This is usually our second bloomer, right behind the anemones. Start looking in early March. They look best at midday. They can also been seen in the late fall when they reappear under damp conditions.

Comments: The Greek name means "false garlic" because the plant has the appearance of an onion but it lacks the onion smell. Unlike most native plants on the Mound, this plant is found throughout Flower Mound, as at the brush arbor at the Flower Mound Presbyterian Church where the entire field is covered at this time. The flower closes during cold or cloudy weather. The False Garlic is considered to be poisonous.

Ground Plum
Astragalus crassicarpus

MAR–JUL

Bean family (Fabaceae):
Buffalo Plum; Pomme-de-Prairie; Milk Vetch; Indian Pea

Description: Our plant is a low, hairy, ground-hugging herb spreading in clumps. The purple-lilac-greenish white flowers are up to 1 inch long, often growing in clusters of 5–25 at the end of the branches. Each flower is two-lipped, with the upper lip standing straight up; the lower lip is three-lobed. The leaves grow to 6 inches in length and are divided into 15–33 small leaflets with rounded tips. The plum-like fruits are inflated round pods that become reddish or purplish when mature. They rest on the ground and look like a nest of small eggs.

Location: The Ground Plum is found mostly on the south-central slope, near the top of the Mound.

Season: The Ground Plum is one of our first arrivals in the spring, arriving in early March. The bright purple flowers will be easy to spot against last season's dry, dead grass.

Comments: The fruits of the Ground Plum are edible and were eaten by both Native Americans and settlers.

Groundsel
Senecio ampullaceus

MAR–MAY

Sunflower or Daisy family (Asteraceae):
Ragwort; Texas Squaw-weed; Texas Butterweed; Clasping Leaf Groundsel

Description: This upright plant grows to 3 feet in height. It has a solitary stem that branches at its upper portion. It is wooly when young to smooth and shiny when mature. The bright yellow flower heads grow to 1 1/4 inches across with (usually) 7–9 narrow, fertile, petal-like ray flowers. The disk-shaped flowers are also yellow. The flowers grow in clusters at the end of a stem, with multiple clusters often forming a flat-topped yellow mass, sometimes 1 foot across. The leaves grow to 4 inches at the base of the plant, and are alternate, wooly, lance-shaped, shallowly toothed, tending to clasp the stem, with one main vein down the center.

Location: This flower is found mainly on the south slope, near the front fence. It favors disturbed areas like the west trail to the top.

Season: It first appears in late March and early April.

Comments: "Senecio" refers to the Latin for "old man." The name Squaw-weed is from its use by Indian women as a health tonic, although it is considered to be poisonous and dangerous to livestock. The name Groundsel comes from its use as a poultice for cleaning wounds. When forming seeds, it creates a round ball of silvery-bristled seeds similar to those of a Common Dandelion. It looks like a ready-made bouquet with many yellow flowering heads on one stem.

Common Dandelion
Taraxacum officinale

Blooms almost all year

Sunflower family (Asteraceae)

Description: The familiar yellow flower head grows to 1 1/2 inches across and is solitary and terminal, growing on a hollow, leafless stalk. It has as many as 200 yellow "tongue-shaped" ray flowers, with no disk flower. The stem produces a milky sap. Its leaves are up to 16 inches long, grouped in a basal rosette. They are oblong, coarsely toothed to deeply lobed, and often widest at the tip. The fruit is the very familiar ball-like arrangement of "parachute-like" seeds that are ready to "fly away" on the wind.

Location: This plant is found in disturbed areas, especially along the northern fence.

Season: Although the Common Dandelion blooms nearly all year, it is most likely to bloom from February to May.

Comments: This is commonly seen as a lawn "weed"; however, it has always been famous as a plant of many medicinal uses and is high in vitamins. The flowers are sometimes used to produce a golden wine. The leaves are edible if gathered wild and early and cooked as greens or tossed in salads, and the roots are also edible. The common name comes from the French *dent de lion*, or tooth of the lion, for the teeth on its leaves.

Southern Dewberry
Rubus trivialis

MAR–APR

Dewberry family (Rosaceae):
Wild Blackberry; Low-bush Raspberry

Description: This is a trailing, scrambling shrub, with hairy to bristly stems growing 3–6 feet in length. These stems often take root at their tips. The branches have prickles. The white and very delicate flowers are essentially solitary. They have 5 petals up to 1 1/2 inches in diameter, with numerous stamens and pistils growing from a dome-shaped center. The compound leaves have 3 to 5 leaflets that grow to 2 inches in length. They are alternate, thick, almost "leathery," elliptic, pointed, evergreen, almost smooth, and sharply/irregularly toothed. The fruits go from green to red to black, maturing in June or July.

Location: There is a large clump of Southern Dewberry near the east fence line, at about its midpoint.

Season: Blooms very early from March to April.

Comments: This Rubus family of plants has over 250 species and produces the raspberry, blackberry, and loganberry fruit. Precise identification is difficult in part because of significant and frequent hybridization. The fruits are an important wildlife food, and humans consume them in the form of delicious pies, cobblers, and jams. Foragers find the plants easily in the spring by the white blossoms and then flag them for picking in early June.

Fringed Puccoon
Lithospermum incisum

MAR–MAY

Forget-Me-Not family (Boraginaceae):
Fringed Gromwell; Narrow Leaf Puccoon; Golden Puccoon

Description: This is an upright, hairy plant growing to 12 inches in height, with several stems growing from a clump (or rosette) of leaves at the base, branching toward the top.
It has numerous, showy, lemon yellow flowers growing about 1 1/2 inches long, with delicate ruffled edges. The trumpet-shaped flowers have 5 united petals. Flowers are numerous on each plant, growing in a curled or coiled terminal spike, which elongates and uncurls as flowers open. It has narrow leaves that grow to 4 inches long, pointed, alternate, hairy, with rolled edges. The basal leaves are sometimes missing or dried up by the time the plant blooms.

Location: The Fringed Puccoon is found primarily on the higher and drier parts of the Mound, especially on the south-eastern slope, but can be found anywhere on the property.

Season: This is an early bloomer, appearing in March before the grasses have turned green.

Comments: The Latin word *incisum* refers to the crinkled edges of this flower. This is a true prairie native plant and is one of the most beautiful found on the Mound. Puccoon is an Indian word meaning "red," referring to the color of dye that can be obtained from the plant's roots. Wool dyers still use this plant as a source for dye. There is contemporary interest in using this plant as a source for modern drugs. If you look carefully, Puccoon can be seen blooming at this time on the roadsides in town, especially on the west end on Shiloh and Wichita Trail.

Poppy Mallow
Callirhoe alcaeoides

MAR–MAY

Mallow family (Malvaceae):
Plains Poppy Mallow; Plains Winecup

Description: The flowers of this plant are borne singly or in branched clusters. They are normally white but occasionally pink and are 1 to 2 inches across. There are five separate petals, the tips of which sometimes appear to be fringed. The flowers are cup-shaped at first (hence "winecup") but open up nearly flat as the plant matures. A distinctive trait of this flower is that the stamens and pistil unite to form a single cone-like structure in the center of the flower. The leaves are alternate. Those at the base of the plant have stems about as long as the leaf. There are few leaves on the upper part of the stem. The leaves are very coarsely lobed or scalloped. Larger leaves have five prominent, deep lobes.

Location: This plant has been seen all over the Mound, near both the north and south fences as well as on the west side. It appears to prefer growth in disturbed areas.

Season: These flowers appear early in the flowering season and last only a few months.

Comments: Our flower is not the common reddish purple variety of Winecup, but usually white or sometimes a delicate lavender. Other members of the Malvaceae family are the hibiscus and the hollyhocks. Okra is the edible fruit of one of the species of hibiscus. Some cotton fibers are derived from this family.

Evening Primrose
Oenothera speciosa

MAR–JUL

Evening Primrose family (Onagraceae):
Pink Lady; Showy Primrose

Description: This often-spreading plant grows to about 18 inches high. The flower has 4 petals, dark rosy pink to white, with darker red veins and a yellow basal spot. They are cup-shaped and up to 2 inches across (less under drought conditions). The petals are initially white when the bud opens but rapidly turn pink. The styles and anthers are a bright yellow. The flowers typically open in late afternoon, persisting until the following morning (and later on cloudy days). The leaves are alternate and lanceolate or oblong, 2–3 inches long, with wavy margins, sometimes toothed.

Location: This colorful flower is found mainly on the southern slope of the Mound, near the fence.

Season: There are scattered early blooms in March, a full bloom in April, and scattered blooms again through summer.

Comments: This is one of our showiest flowers on the Mound, frequently colonizing extensive areas. The flower has ultraviolet markings that cannot be seen by the human eye but are readily visible to bees' eyes.

Texas Dandelion
Pyrrhopappus multicaulis or P. carolinianus

APR–JUN

Sunflower or Daisy family (Asteraceae):
Many Stemmed False Dandelion; Pata de Leon

Description: The plant grows up to 20 inches tall. The stems grow singly or in groups from the base, and they branch erratically. Stems exude a milky substance when bent or broken. The flowers are up to 2 inches across and grow singly or in a small group at the end of short stalks. There are many lemon yellow ray flowers growing in two or three rows on each flower head. There are no disk flowers. Leaves grow up to 6 inches long, are alternate and stalked, mostly at the base. They are usually deeply lobed but sometimes only mildly toothed, especially toward the top of the stem.

Location: This flower is located mainly on the south slope of the Mound.

Season: Seen as early as the first of April and still around as late as June or early July.

Comments: The flowers open in the morning and close around midday. The young leaves are often gathered and used with other salad greens. Native Americans enjoyed their roots. The numerous fruits of this plant are tipped with a spreading tuft of feathery bristle, typical of most dandelions.

Texas Bluebonnet
Lupinus texensis

MAR–MAY

Legume, Bean or Pulse family (Fabaceae):
Buffalo Clover; Texas Lupine

Description: This is an upright, sprawling annual that grows up to 2 feet tall. Very fragrant, each individual flower is small (1/4 inch) with five deep blue petals. Each flower has a white (or yellowish) spot (turning purplish after pollination) at the base. Individual flowers cluster on the upper 2–6 inches of the stem in a dense terminal spike with a pointed, conspicuously silvery-white tip. Leaves are palmate, with 4–7 (usually 5) pointed leaf segments radiating from a central point, on a long stem. Each plant produces one to several bean-like hairy fruits that are 1–2 inches long.

Location: This beautiful plant can be found on the south slope, mainly near the south fence line.

Season: One of the early bloomers in March.

Comments: The Texas Bluebonnet is one of six native Lupinus species that constitute the state flowers of Texas. All of the bluebonnets on the Mound have been planted here, as this is a Hill Country plant and doesn't grow this far north naturally. The species in Big Bend grows to 3 feet tall. The bean pods explode to disperse the seeds. After the blooms fall and the seeds are mature, they can be pulled up and scattered on a new site where they will disperse and make new plants the next spring.

Celestial
Nemastylis geminiflora

MAR–MAY

Iris family (Iridaceae):
Prairie Celestial; Prairie Pleatleaf; Prairie Iris; Celestial Ghost Iris

Description: These beautiful and striking flowers grow up to 2 inches across and are a pale blue-violet, sometimes lightening to white "eyes" at the base. The 6 petals are somewhat short and lance-like, ending in a point. The prominent orange anthers are erect in the morning but droop as the day lengthens. The plant grows to about 16 inches in height, with few or no branches. The leaves are a bright green, alternate, and quite long and slender (grass-like).

Location: This plant is found throughout the Mound, mainly in the southeast corner and near the top on the southern slope and the west fence.

Season: An early bloomer, the best show is in April.

Comments: Individual flowers are open for only a few hours, opening in late morning and usually closing by about 3:00 p.m. Each new bloom lasts only about one day. Without a bloom, the Celestial is difficult to find because its leaves are grass-like and blend in with the surrounding Bluestem. The flowers often appear in pairs, hence *gemini* (twin) *flora* (flower).

Prairie Verbena
Glandularia bipinnatifida

MAR–OCT

Vervain or Verbena family (Verbenaceae):
Dakota Vervain; Small-flowered Vervain; Western Pink Vervain

Description: The plant grows to about 18 inches in height. Individual flowers are small (up to 3/8 inch across), varying in color from purple to lavender to pink. They are trumpet-shaped, with 5 united petals and 5 sepals. Flowers are numerous, growing in rounded, terminal clusters. The stems are not round, but four-angled, branching from the base. The lower branches sprawl or creep along the ground and take root wherever a joint touches, forming dense colonies. The leaves vary up to 3 1/2 inches in length, are opposite and stalked. Each leaf blade is deeply cut or lobed into numerous narrow segments and is hairy.

Location: The plant is found all over the Mound, particularly on the southern slope of the hill, toward the top.

Season: The plant blooms most freely in the spring, but if the conditions are moist, it will bloom most of the year.

Comments: The Prairie Verbena is one of the most abundant and familiar wildflowers of Texas. It is an important nectar plant for butterflies.

Texas Star
Lindheimera texana

APR–MAY (later with rains)

Sunflower or Daisy family (Asteraceae):
Texas Yellow Star; Star Daisy; Lindheimer's Daisy

Description: The plant grows up to 20 inches tall. It may begin flowering when it is not over 2 inches tall and will continue doing so to maturity. The plant itself is somewhat hairy with a leafy stem that often branches toward the top. The flower head grows up to 1 1/4 inches across. There is one to several flower heads in a cluster at the end of each stem. Each flower has 5 bright yellow ray flowers (petals), notched at each tip. The disk flower (center) is yellow as well. There are many leaves, growing up to 3 1/4 inches long. They are alternate and coarsely toothed at the base, crowded on the stem, and tapered at both ends.

Location: This plant is found primarily near the crest of the Mound, especially on the south slope.

Season: First showing around the middle of April, this pretty flower is over by the end of May.

Comments: This variety is named for Ferdinand Lindheimer (1801–1879), who was a German botanist who settled and did research in New Braunfels, Texas. The plant is more common to the Hill Country than this area.

Wild Hyacinth
Camassia scilloides

APR

Lily family (Liliaceae):
Eastern Camas; Meadow Hyacinth; Meadow Quill; Camas Lily

Description: The plant grows up to 2 feet in height. Each flower is about 1 inch across, lavender to pale blue, and sweet-scented. Each flower has 6 petal-like growths (3 petals; 3 sepals). The flowers are numerous on each plant, being loosely congested on elongated terminal clusters on a leafless stem. The leaves grow up to 16 inches in length and are about 1/2 inch wide and are alternate on their stem. The midrib of each leaf is prominently ridged.

Location: This plant is found primarily in the gilgai (hog wallows), clustered on the northwest slope, about midway to the north fence. They enjoy the extra water trapped there.

Season: This flower blooms briefly and dramatically in early April and is gone by the end of the month.

Comments: The bulb is edible and was an important food source for Native Americans and early settlers. The Native method of cooking in a pit with hot stones is preferred as it is said to bring out their natural sugars. There is an account in the Flower Mound history of a man who prepared them as a boy. Archaeologists look for this plant in the spring as likely sites to search for Native tools lost during digging of these bulbs. This endeavor should not be undertaken by the inexperienced, as Wild Hyacinth closely resembles the Death Camas, a deadly poisonous plant. Death Camas has been observed in the LBJ National Grasslands in Decatur.

Ratany
Krameria lanceolata

MAY–OCT

Ratany or Krameria family (Krameriaceae):
Trailing Ratany; Crameria; Prairie Bur; Sand Bur

Description: The plant is quite slender, intricately branched, and vine-like, extending 2 feet or more. The flower is quite small (less than 3/4 inch across), reddish-wine colored, consisting of 4–5 petal-like sepals of the same color as the 5 very unequal and smaller petals. The upper 3 petals are colored and united (fan-shaped); the other 2 are greenish and short. Each flower is solitary on a short hairy stalk. The leaves grow to 1 inch and are long and narrow, alternate, and densely covered with silky hairs.

Location: This tiny flower is usually found "buried" under taller plants and grasses. It is sparsely distributed all over the Mound.

Season: This plant is flowering by May, but it is so small it may be overlooked. It continues to bloom all summer.

Comments: This plant is believed to be somewhat parasitic, with its roots drawing nutrients from the roots of other plants. Since the vine hugs the ground, the tiny flowers are often covered by other plants and are difficult to see. Look for the deep violet color. Native Americans used this plant medicinally, for tanning leather, and as a dye source. The name "Sand Bur" comes from the exceedingly painful sticker this plant produces. Large like a Goatshead, its seed head is the way the plant disperses.

Philadelphia Fleabane
Erigeron philadelphicus

MAR–MAY (plus fall)

Sunflower family (Asteraceae):
Common Fleabane; Fleabane Daisy; Philadelphia Daisy

Description: This upright, slender, hairy plant grows up to 28 inches tall. There is usually a solitary, grooved stem, branching in the upper portion. The flower head is up to 1 inch across with a very large yellow-orange disk flower surrounded by 150 or more tiny (threadlike), white (occasionally light pink) ray flowers. (The ray flowers are much finer than on other daisies.) The flower heads themselves form a large, loose terminal cluster at the end of the slender branches. The spatula-like leaves are up to 6 inches long and 1 inch wide. The lower leaves are long-stalked and toothed or lobed; the upper leaves are stalkless and tend to clasp the stem at the base of the blade.

Location: This plant is found all over the Mound, especially on the slopes, near the crest.

Season: It blooms early and, with sufficient moisture, again in the fall.

Comments: Dried and crushed parts of the plant have been used in teas for sore throat and stomach ailments. Before insecticides, this plant was used to repel fleas (hence "Fleabane").

Bluet
Hedyotis nigricans

MAR–OCT

Coffee or Madder family (Rubiaceae):
Star Violet; Fine Leaf Bluet; Prairie Bluet

Description: This plant grows to 20 inches in height. Its stems are slender and solitary or a few form a common base, with significant branching on the upper portion. The tiny flowers, few to numerous, grow to 1/4 inch across and have 4 white to lavender-pink petals united at the base (trumpet-shaped). The petal tips are sharply pointed in the shape of a Maltese cross and form small, leafy, crowded clusters above the foliage. The leaves grow to 1 1/2 inches long and are very narrow (threadlike), opposite, short-stalked or stalkless, with margins that are often rolled toward the lower surface.

Location: This plant is found in small clusters just about everywhere on the Mound.

Season: The Bluet is an early harbinger of spring, sometimes blooming as early as February. It can continue to bloom into the fall if the moisture conditions are good.

Comments: The plant is called the *Hedyotis nigricans* because the leaves turn black when they die.

Mexican Hat
Ratibida columnaris

MAR–NOV

Sunflower family (Asteraceae):
Thimble Flower; Prairie Coneflower; Red Spike Mexican Hat

Description: The plant grows up to 2 1/2 feet in height, with one to several grooved stems growing from the base. The flower head is very distinctive, growing singly on long, slender branches. There are 4–10 drooping, petal-like ray flowers, up to 1 inch long and varying in color from yellow to solid reddish-brown or, more commonly, velvety reddish-brown at the basal portion and yellow at the tip portion. You will find many differently colored varieties of the Mexican Hat on the Mound. The center (disk flower) portion of the flower is green or brown and is an erect and conspicuously elongated (up to 2 1/2 inches) cone. The leaves are 2–4 inches long and half as wide, alternate, with the blade divided into 5–11 narrow segments, which can also be further divided. The top portion of the flower-bearing stem is bare of leaves.

Location: This flower is found all over the Mound, particularly on the southern slope.

Season: This is one of the longest blooming plants found on the Mound, visible from March to November.

Comments: This distinctively shaped flower resembles the classic sombrero or "Mexican hat," hence its name. It looks similar to the Clasping Leaved Coneflower except that the leaves of the two species are quite distinct from each other. The flower is wide ranging and is found from the East Texas woods to the far Pecos and Panhandle areas.

Blue-eyed Grass
Sisyrinchium pruinosum

APR–MAY

Iris family (Iridaceae):
Dotted Blue-eyed Grass; Grass Violet

Description: The plant grows up to 12 inches in height and is usually seen in clumps. The flowers are small (dime-sized) with 6 violet-blue-purple petal-like segments with a yellow center or "eye." (Actually, there are 3 petals and 3 sepals that are similar in both shape and color.) Flowers can appear in a small cluster at the tip of the stem. The flowers tend to close in the afternoon. The stems are flat or somewhat winged, resembling grass-like leaves. The leaves are long (up to 9 inches), quite slender (grass-like), alternate, flat, growing from the base, and tend to sheath the stem.

Location: This plant is evident all over the Mound.

Season: Blooms in April and May.

Comments: There are numerous species of the *Sisyrinchium*, and it is often difficult to distinguish one from the other, with distinctions based on such things as branching pattern and leaf length. This plant is common in yards in Flower Mound. Some Indians used it as a medicinal tea.

Color Chart
February–March

White
- Ten-Petal Anemone
- False Garlic
- Southern Dewberry
- Poppy Mallow
- Evening Primrose
- Philadelphia Fleabane

Yellow
- Bladderpod
- Groundsel
- Common Dandelion
- Fringed Puccoon
- Texas Dandelion
- Texas Star

Blue
- Ground Plum
- Texas Bluebonnet
- Celestial
- Prairie Verbena
- Wild Hyacinth
- Bluet
- Blue-eyed Grass

Red
- Ratany
- Mexican Hat

Chapter 6
Explosion of Color
The Flowers of April and May

While this section only represents two months, it is the period of greatest bloom activity on the Mound. By now, the Mound is entirely green, and grasses and flowers are in fierce competition to dominate resources. Almost every day, a new species pops up and changes the relationships. Some will bloom for a long period, some for only a day.

Wild Foxglove

Wild Garlic
Allium drummondii

APR–MAY

Onion, Garlic, and Leek family (Liliaceae):

Prairie Onion; Wild Onion; Drummond's Wild Onion

Description: This is an upright to somewhat sprawling, smooth plant growing to 12 inches tall. The flowers are small (to 3/8 inch wide), white to pink (occasionally lavender to purple-red), fading as they mature, with 6 petal-like segments and 6 stamens, bell-shaped, in loose clusters of 10–25 flowers at the top of the hollow, leafless stem. There are 3 or more long (to 12 inches), slender leaves, conspicuously creased along the midrib, growing only from the base.

Location: This plant is common all over the Mound, especially in unshaded areas in the front and in moist areas.

Season: Early bloomers appear in late April and continue through May. They may reappear again later in the season if moisture conditions are high.

Comments: *Allium* is the Latin name for garlic. This edible plant has a strong onion smell, as do other members of this *Allium* species. A confusing look-alike is the False Garlic (Crow Poison), which appears at the same time but lacks the distinctive odor of the onion. The Indians and early settlers used the Wild Garlic bulbs for food. The name "Chicago" is reported to derive from the Indian word for this onion. The plant has also been used to treat burns, wounds, and bee stings.

Yarrow
Achillea millefolium

APR–JUN

Sunflower, Daisy, or Aster family (Asteraceae):
Milfoil; Common Yarrow

Description: This is a stiffly upright, hairy plant growing to 40 inches tall, usually unbranched except near the flowering layer. The plant forms clumps or small colonies. Each tiny flower (to 1/4 inch across) has 4–6 yellowish-white petal-like ray flowers. The central disk flower is white to pink. The flower head consists of 10–20 flowers, forming a dense cluster with several clusters forming a terminal mass on the same plant. It has soft, gray-green, aromatic leaves that are stalkless, alternate, growing to 6 inches in length, very finely divided into thread-like segments. The leaves have an extremely delicate, fern-like, lacy appearance.

Location: When in bloom, this plant can be found all over the Mound.

Season: It first appears in April and lasts until June.

Comments: The Yarrow was brought from Europe where it has been used medicinally since early history to stop blood flow. It has also been made into tea to treat influenza, gout, and kidney ailments. It is frequently mistaken for "Queen Anne's Lace." The Latin *millefolium* means "thousand leaves," by which it is easily identified.

Ohio Spiderwort
Tradescantia ohiensis

APR–MAY

Spiderwort (Dayflower) family (Commelinaceae)

Description: The plant grows up to 3 feet tall. It has a gluey sap and is covered by a white powdery coating that comes off when rubbed. The flower can vary in color from white to a dark blue or rose. Each flower is up to 2 inches across, with 3 equally sized petals, 3 sepals, and 6 prominent yellow-tipped stamens. They grow in clusters on the stem, and the plants themselves tend to grow in clumps. The leaves are quite long (up to 18 inches), narrow, pointed, alternate, and stalkless. They clasp the stem at the base.

Location: There is a large colony of this plant along the east fence, especially toward the north end, where they grow very large. There is a smaller variety along the west fence near Tom Thumb.

Season: The season is short for this flower, both within the year and within the day. Flowering begins in April and is over in May.

Comments: The flowers are very delicate and only bloom for a few hours in the morning. This is the most widespread spiderwort in the United States. It is common along railroads in Denton County. "Wort" is an Old English word for herb.

Prairie Phlox
Phlox pilosa

APR–MAY

Phlox or Polemonium family (Polemoniaceae):
Downy Phlox; Sweet William

Description: This plant grows up to 20 inches tall. The flower normally has 5 petals, varying in color from pale rose to bluish lavender (occasionally white). The flowers cluster on the end of the stem. There are 5 stamens of varying length. The leaves are either alternate (toward the top) or opposite (toward the bottom) of the stem. Their blades are narrow and very pointed.

Location: Scattered throughout the Mound, the largest group of the Prairie Phlox appears on the northwestern slope, between the top of the Mound and the fence line.

Season: The season for Phlox is relatively long; they come in around the first of April and continue sporadically until late in the year.

Comments: There are more than 300 species of Phlox. Accordingly, it is sometimes difficult to distinguish one from the other. The flower emits an exceptionally pleasant fragrance. Our Phlox comes in many variations of shape and color. It is very possible there is more than one species on the Mound.

Wild Foxglove
Penstemon cobaea

APR–JUN

Foxglove, Figwort, or Snapdragon family (Scrophulariaceae):
Foxglove; False Foxglove; Balmony; Dewflower

Description: This hairy plant grows to 2 feet tall, sometimes in clumps or colonies. The bell-shaped flower is up to 2 1/4 inches long, usually pink tinged with reddish-purple or lavender and streaked inside with dark purple lines. It is broadly open and five-lobed at its rim. The lower lip of the flower has no hair on its surface. The flower has 5 stamens, of which one bears no pollen but is bearded with yellow hairs along its length. The individual flowers tend to be grouped in showy clusters around the downy stem, forming a terminal spike. The downy, clasping, thick, sharply toothed leaves grow up to 3 1/2 inches long, becoming smaller (and stalkless) toward the summit of the stem.

Location: This beautiful flower is found circling the top of the Mound about 20 yards down. It is one of the showiest and most dramatic of the Mound's flowers.

Season: Emerging in April, the full show is around the first of May.

Comments: This plant is named for a Jesuit priest, Father Bernardo Cobo, because of its similarity to exotic tropical American flowers. Early settlers brewed tea from its leaves for a laxative. Today, the leaves of another foxglove are used in the preparation of digitalis, a medication for treating heart disorders.

Indian Paintbrush
Castilleja indivisa

APR–JUN

Foxglove, Figwort, or Snapdragon family (Scrophulariaceae):
Texas Paintbrush; Entire-leaf Paintbrush; Scarlet Paintbrush

Description: The plant can grow to heights of about 16 inches. The stems are often purple and hairy, with several growing from a common base, forming clumps. The red-tipped (occasionally light yellow) flowers grow to 1 inch across and are surrounded by similarly colored floral leaves called bracts. The flowers and their bracts cluster, forming a very showy terminal spike. The leaves grow up to 4 inches long, are alternate, spikeless, and narrow.

Location: Samples of this showy plant may be found all over the Mound, singly or in small clumps.

Season: April to June (sporadically until the fall).

Comments: It gets its name because of a resemblance to a brush dipped in paint. The plant is semiparasitic. It is believed to be poisonous to animals; however, the Native Americans used it medicinally. The yellow Paintbrushes sometimes seen with the red flowers are variations, not the separate species Lemon Paintbrush, *Castilleja citron*, which more closely resembles the Purple Paintbrush.

Variations: *Castilleja purpurea*, Purple Paintbrush: One of the prettiest of the Paintbrushes, the purple variety has many different colors and hues. We once had one of these plants, near the east trail. Noted on the way up before a sunrise service, it was discovered missing on the way down and has not been seen since. The strongest argument for not picking the flowers on the Mound is that you might pick the last specimen.

Sensitive Briar
Mimosa roemeriana

MAY–JUN

Legume, Bean, or Pulse family (Fabaceae):
Roemer's Sensitive Briar; Catclaw

Description: This is a branching, trailing plant with short, curved spines and round balls of pink flowers on long stalks arising from the axils of compound leaves. The individual flowers are quite tiny, crowding in very dense, round clusters about 1 inch across at the tip of a leafless stalk. The leaves are up to 6 inches long, divided into 4–8 pairs of small segments, with these again divided into 8–15 pairs of smaller leaflets. The leaf-stalks themselves are prickly. The seed pods are a curious, flat, spiny, three-pronged growth with each of the 3 arms being about 3 inches long, changing in color from green to reddish-brown as they mature.

Location: This plant is dispersed throughout the Mound.

Season: In full bloom by May, this exotic flower is well worth pursuing; by June, most of the blooms have converted into long spiky fruits.

Comments: The leaves are touch-sensitive, folding when disturbed. The leaves also close at night or in cloudy weather. Children delight in this plant; it is fairly easy to find, even when it isn't blooming, by the mimosa-like leaves. The flowers of this plant are shaped similarly to those of the Prairie Acacia, except that the flowers of the latter are white.

Yucca
Yucca arkansana

APR–MAY

Yucca, Century Plant, or Agave family (Agavaceae):
Soapweed; Spanish Bayonet

Description: The leaves of the Yucca are "spear-like," very pointed, quite stout, narrow, growing mainly from the base of the plant and radiating around the central stem. They grow to more than 2 feet in length. The numerous ivory or white bell-shaped flowers have 6 petal-like segments and form a cluster more than 2 feet long on the upper part of the stem. Each bloom is about the size of a billiard ball.

Location: There are several colonies of Yucca plants on the Mound. The largest groups are located on the north and south slopes, just below the top of the Mound.

Season: The Yucca plant itself is an evergreen. It shoots up tall, dramatic stalks in April that go to full flower in May.

Comments: There are many varieties of Yucca. The Yucca ranks foremost among the wild plants used by the Native Americans. It has served as a source for food, fiber, soap, and medicine. Ropes, sandals, mats, and baskets have been made from the leaf fibers. The spiny tip was broken off, leaving a long fiber attached, and used as a sewing needle. The root provides an effective shampoo, being high in saponin, a natural soap. It is pollinated only by the Yucca moth. The flowers are the edible portions. Lewis and Clark referred to this plant as Beargrass.

Indian Plantain
Arnoglossum (formerly Cacalia) plantagineum

APR–JUN

Sunflower family (Asteraceae):
Finned Indian Plantain; Prairie Indian Plantain; Prairie Plantain

Description: This plant has a tall, thick, solitary, grooved, hairless, often dark purple stem growing to 3 feet high. When near the flowering portion, the stem branches, yielding numerous clusters up to 6 inches across of small, green to creamy white, upright, ribbed flower heads about 1/2 inch tall. Each flower head looks like an unopened, angular bud. Closer analysis reveals 5 petals and 5 lobes. Together, they form multiple large, loose, rather flat-topped terminal masses. The extremely large leaves, on a long stem, grow to 8 inches in length and 3 inches wide. They are alternate, oval-shaped, thick, and rubbery/fleshy, with numerous (7–9) prominent longitudinal veins converging toward the tip. Most of the leaves are clustered near the base of the plant, with relatively few leaves on the stem.

Location: This plant is located all over the Mound, with a large colony on the northeastern slope.

Season: It is usually in full bloom by mid-May. It may reappear in the fall if there is abundant rainfall.

Comments: The "plantain" name comes from the leaf's resemblance to the banana-like plantain leaf. The Indians used the plant as a poultice for cuts, bruises, and infections. It is also found in the northern prairies all the way to Canada.

Purple Horsemint
Monarda citriodora

MAY–SEP

Mint family (Lamiaceae):

Plains Beebalm; Lemon Mint; Plains Horsemint

Description: The hairy plant grows to 2 feet in height. The stem is square, not round, and is usually solitary, except that it sometimes branches toward the top. The individual flowers are small (up to 3/4 inch tall) with 5 petals, but they form a prominent, elongated, terminal spike cluster of 2–6 successively dense, terminal heads (head-like whorls) and have large, sharp-tipped colored bracts immediately below the various clusters that are unlike the petals. The flower petal colors vary from white to pink or lavender, often with purple dots. The leaves grow up to 3 inches in length, are opposite and long-stalked, some with a few teeth on the margin.

Location: This plant can be found all over the Mound, mostly on the southwestern and south central slope, toward the top.

Season: While this beebalm can be seen flowering from May to September, the best bloom period is in early June.

Comments: The flower has a lemon scent. Native Americans used it as a perfume and insect repellant. The dried plant is quite prominent in the fall, once the colors are gone, when it has an unusual two- three-layered dried bract growth on the same stem. Our flowers are white, sometimes with spots; in Palo Pinto County, the flowers are purple. Tea can be made from the leaves of this species; an even better tea is brewed from relative *Monarda fistulosa*, Wild Bergamot. This is the tea used during the Revolution after the Boston Tea Party when English tea was boycotted. Many herbs for cooking come from the mint family.

Prairie Larkspur
Delphinium virescens

MAY–JUN

Buttercup or Crowfoot family (Ranunculaceae):
Rabbit Face; Plains Delphinium; White Larkspur

Description: This is an upright, very slender, hairy plant growing up to 3 feet in height. The flower is whitish, crinkled, with 5 petal-like sepals and 4 smaller petals. The uppermost sepal is drawn backward into a "spur." The lower inner petals are covered with long hairs. The petals are clustered at the center of the flower. The flowers tend to cluster on a long (up to 20 inches), narrow, finely downy terminal spike that grows well above the foliage. The alternate leaves are about 3 inches long and equally as wide, clustered toward the base of the plant. Leaf segments are heavily forked.

Location: This striking plant can be found most abundantly near the south fence. Later in the season, it can be seen in all portions of the Mound.

Season: The first of May is the early emergence of the Larkspur, which continues dramatically throughout the month.

Comments: The Larkspur kills almost as many cattle as Locoweed and is considered poisonous. The name Larkspur refers to a resemblance to the hind spur on a European Lark. The Latin name refers to the flower's resemblance to a dolphin.

Variations: *Delphinium carolinianum*, Wild Blue Larkspur, Carolina Larkspur. A beautiful, blue variety has been seen outside the fence but not inside. It may be a roadside invader.

Carolina Larkspur

Texas Vervain
Verbena halei

APR–OCT

Vervain or Verbena family (Verbenaceae):
Slender Vervain; Candelabra Vervain; Blue Vervain

Description: This is an upright, smooth, delicate, slender-branched plant growing to 2 1/2 feet in height. The stem may be solitary or several growing from a common base, four-angled, grooved, and branched at its upper portion. The flower is tiny (1/4 inch across), bluish lavender or pink, trumpet-shaped, with 5 petals. The flowers on each plant are numerous, forming long, very slender spikes at the top of the plant. The leaves are opposite and grow to 3 inches in length, with the longer blades deeply lobed.

Location: This plant appears in isolated groups all over the Mound.

Season: It flowers most prolifically from May to June and sporadically afterward until about October.

Comments: Vervain is a very historic medicinal, going back to the Roman times when it was used to treat eye disorders.

Scarlet Gaura
Gaura coccinea

APR–MAY

Butterfly Weed or Evening Primrose (Onagraceae):
Willow Herb Gaura; Scarlet Bee-Blossom

Description: This hairy plant grows, singly or in clumps, to 3 feet in height. The white or pink, very fragrant flower is about 1/2 inch across. It has 4 asymmetrically arranged petals that spread upward. There are 8 long and very conspicuous stamens, pointing downward. The flowers are very numerous on a slender spike (raceme) at the end of a leafless stalk. The very narrow leaves grow to 2 1/2 inches in length and 1/2 inch wide. They are alternate, stalkless, and crowded on the stem.

Location: Only seen in one large colony on the southern slope of the Mound, about halfway up the west trail, on the right.

Season: Best seen in April as they first flower; they then quickly degrade, losing the white portion of the flowers.

Comments: These flowers are common on the roadsides and in highway department plantings, but they are best viewed up close.

Old Plainsman
Hymenopappus scabiosaeus

APR–JUN

Sunflower family (Asteraceae):
Wooly White

Description: The plant grows up to 3 feet in height on an erect, stout stem that is branched only toward the top. The stems have white, felt-like hairs. The individual flowers are tiny and funnel-shaped with a long, prominent stamen protruding from each one. White or creamy white flowers are found at the end of the stems, sometimes having as many as 60 of these tiny flowers in a ball-like cluster. A defining characteristic of this plant is the "lacy" leaves that are up to 5 inches long, grayish-green, and divided into innumerable fern-like hairs.

Location: This fairly prominent plant is randomly dispersed all over the Mound, with perhaps more on the south slope than elsewhere.

Season: Emerging in April and enduring through June and July.

Comments: Common in unmowed fields in Flower Mound, it can easily be spotted from the road. This is another plant used medicinally by Native Americans.

Variations: Fine Leaf Wooly White, *Hymenopappus artemisiifolius*. The main difference in this form and the Old Plainsman is the leaf structure.

Prairie Parsley
Polytaenia nuttallii

APR–JUN

Carrot or Parsley family (Apiaceae):
Wild Dill; Texas Parsley; Prairie Parsnip

Description: This is a stiffly upright, smooth-stemmed, stout plant growing up to 3 feet in height. It has a solitary stem that branches in the upper portion. Each yellow flower is tiny (1/4 inch or less across) and has 5 petals. The flowers grow in ball-like clusters up to 2 inches across, with perhaps 10–20 clusters per stem, on stalks of unequal length. The leaves are large (up to 7 inches long by 6 inches wide), each divided into broad, deeply cut segments, with scalloped, lobed, or coarsely toothed margins.

Location: This plant grows singly or in small clusters all over the Mound, especially on the southeastern slope.

Season: A long blooming season helps to make this common prairie plant familiar.

Comments: The leaves are sometimes used in cooking much as cultivated dill. The stout stem, though dried out, remains upright even through the winter. It is common in native prairies.

Venus' Looking-glass
Triodanis perfoliata

APR–JUN

Bluebell family (Campanulaceae):
Hen and Chickens

Description: This stout, hairy plant grows up to 12 inches in height. It has a solitary, angled, leafy stem, mostly unbranched. The flowers are up to 3/4 inch across, purplish to bluish-violet, and deeply five-lobed. They are fairly flat when fully open and tend to grow from the junction of the leaves and the stem. There are several flowers forming a cluster along the stem. The lower flowers on the stem often do not open (but they do produce seed). The rounded hairy leaves are up to 1 inch across, alternate, and deeply notched at the base where they clasp the stem.

Location: This flower is found sparsely all over the Mound, especially on the southern slope about halfway to the crest.

Season: Blooming from April to June, this small plant requires a careful search.

Comments: The Cherokee Indians used the root of this plant, in combination with other herbs, to make a drink to address problems with indigestion. These flowers are capable of self-pollination.

Yellow Sweet Clover
Melilotus indicus

APR–MAY

Legume or Bean family (Fabaceae):
Sour Clover; Alfalfilla; Indian Clover

Description: The plant grows up to 4 feet in height. The individual yellow flowers are quite small (about 1/4 inch long), with 5 petals. They form slender, cylindrical, spike-like clusters, 2–6 inches long, along numerous branches, which grow from the same point on the stem as the leaf. The leaves are alternate and are divided into three finely toothed leaflets.

Location: This plant is found mainly near the summit of the Mound, toward the western slope, as well as on the east side, near the fence line.

Season: Find this one blooming in April and May.

Comments: Sweet clover can be poisonous to cattle. All *Melilotus* are invaders and not native and are removed where possible in native prairies. The plant is now used to formulate a medically used anticoagulant and is cultivated to enrich the soil with nitrogen. The leaves, when dried, have a fragrance of vanilla and are often used in making sachets.

Western Horsenettle
Solanum dimidiatum

APR–SEP

Nightshade family (Solanaceae):
Potato Weed

Description: This prickly plant grows up to 3 feet in height. Each plant has one or more flowers with 5 blue-purple petals that are united at the base to form a flat, five-pointed star-shaped flower up to 1 1/4 inches across. It has 5 bright yellow, prominent, long and thick protruding stamens. The somewhat prickly (underneath) leaves grow to about 4 inches in length and are often conspicuously lobed, dark green, and somewhat hairy underneath.

Location: This plant is located all over the Mound.

Season: This common plant has a long flowering season. Its fruit easily identifies it.

Comments: The plant, particularly the small (1/2 inch in diameter), yellow, tomato-shaped fruit, is poisonous (even lethal) to livestock. The fruit remains on the plant for many months. It is suspected of causing birth defects and abortions in livestock. It is the cause of the neurological disease known as "Crazy Cow Syndrome," wherein the victim becomes disoriented, uncoordinated, and accident-prone. This plant is related to both the potato and the eggplant. The eggplant flowers in your garden are virtually identical.

Variation: Silverleaf Nightshade or White Horsenettle, *Solanum elaeagnifolium*. The flower of this plant appears to be identical to that of the Western Horsenettle. However, the leaves of the Silverleaf Nightshade grow to 4 inches in length and are relatively unlobed, perhaps with a strongly wavy margin, are silvery to gray, with an underside that is without prickles and somewhat velvety in appearance.

Locoweed
Oxytropis lambertii

APR–JUL

Legume, Bean, or Pulse family (Fabaceae):
Purple Loco; Lambert's Crazyweed; Stemless Loco

Description: The Locoweed is an upright, hairy plant growing in tufts or clumps up to 14 inches high. The plants often form dense colonies from underground runners. Each pea-like flower is 1/2–1 inch long, variously purple, red, lavender, or rose in color and quite fragrant. The flower has 5 petals. Flowers grow in clusters of 10–25 on a stem, forming dense, elongated terminal racemes on leafless stalks. The leaves are up to 12 inches long, clustered at the base of the plant. Each leaf is divided into 9–19 leaflets, each of which is up to 1 1/2 inches long.

Location: Very rarely seen near the top of the Mound.

Season: April is the best chance to see this rare one.

Comments: The *Oxytropis* is a genus of about 300 species worldwide, some of which are highly toxic and harmful to livestock. Horses are particularly susceptible to this Locoweed, with symptoms varying from paralysis to death. All parts of the plant are toxic. This is the flower that Edward Marcus is holding in the photograph with Lady Bird Johnson. She left it with Bob Rheudasil for identification.

Edward Marcus with Lady Bird Johnson on a trip to the Mound in the early 1970s. Mr. Marcus holds a sample of a Locoweed later identified for Mrs. Johnson. Photo courtesy of Black Mark Farms.

Engelmann Daisy
Engelmannia pinnatifida

APR–JUL

Sunflower family (Asteraceae):
Cut Leaf Daisy; Engelmann's Sunflower

Description: This is an upright to spreading, stout-stemmed, densely coarse-hairy plant growing up to 3 feet in height. The stem branches toward the upward portion to form a rounded crown with numerous flower heads in terminal clusters on the stem. The long-stalked, yellow flower head is up to 2 inches across. It normally has 8 ray flowers (petals) that are up to 1/2 inch long, indented at the tip, and that expand outward during the early part of the day. They droop or fold downward in the heat of the day. The center of the flower head (the disk flower) is also yellow. The alternate and very hairy leaves grow up to 12 inches long and are found mostly toward the base of the plant. They are deeply cut or lobed, with the lobes themselves being further lobed or toothed.

Location: This plant is found mainly toward the top of the Mound, on the east and southern slopes.

Season: Blooming in April and continuing through July.

Comments: It appears to be a preferred food for livestock. Accordingly, it is now overgrazed in most farm locations. It is named after Dr. George Engelmann, a St. Louis botanist, who supported the work of Ferdinand Lindheimer, the "Father of Texas Botany."

Goatsbeard
Tragopogon dubius

MAY–JUL

Sunflower family (Asteraceae):
Noon-Flower; Western Salsify; Yellow Salsify

Description: This plant grows up to 32 inches in height. The unique looking, bright yellow flower head is up to 2 1/2 inches across, with 100–200 petal-like central ray flowers with a prominent set of approximately 13 protruding bracts under each flower. There is no central disk flower. The flower heads are solitary on the end of the branch. Generally speaking, they are open only during morning hours. Each plant has numerous hollow, upright stems with few branches growing from a common base, forming a bushy clump. The stems are difficult to break and contain a milky sap. The leaves are grass-like, stemless, pointed, and are up to 20 inches long. They clasp the branch.

Location: This plant appears mainly along the east fence line, especially near the midpoint, and in the southeast corner of the Mound.

Season: Blooms from late spring through summer.

Comments: The pappus or "fruit" of this plant is a giant (about softball size), feathery, ball-shaped arrangement of seed-bearing "parachutes," similar to the dandelion. The basal leaves can be eaten in salads or as cooked greens. All the species in America are invaders from the Eurasian region.

Blue Sage
Salvia azurea, var. grandiflora

MAY–NOV

Mint family (Lamiaceae):
Big Blue Sage

Description: This is an upright, widely sprawling, hairy plant growing up to 5 feet in height. The stem is solitary to a few growing from a common base. It is four-angled and branched in its upper portion. The 1-inch long, purple to blue to whitish flower has 5 petals—2 on top and 3 on bottom—forming a lip. The lower lip has 3 lobes (middle lobe notched) and has 2 prominent white areas toward the base of the lower lip. The flowers usually grow in several small clusters that form a long, slender terminal spike up to 7 inches long. The leaves grow up to 4 inches long, are opposite, stalked, lanceolate, with margins entire or slightly toothed.

Location: This plant can be found virtually all over the Mound in the fall.

Season: The Blue Sage becomes more prominent into the fall as other flowers disappear. In the October–November time frame, the Blue Sage and the Goldenrod are the most prominent flowers on the Mound.

Comments: Our plants flower in a wide range of blues to lavender and almost white. The flower is designed to promote cross-pollination. The firm upper lip serves to protect the nectar from dilution by rain. The larger lower lip serves to attract and support insects. Some of this species are used medicinally, and some are herbs. Some of the seeds are rich in protein and were once used as food.

Yellow Ground Cherry
Physalis pubescens

APR–DEC

Nightshade family (Solanaceae):
Downy Ground Cherry; Low Hair Ground Cherry

Description: This is an upright, usually hairy plant growing to 3 feet in height. The drooping flower is 3/8 inch long, 5/8 inch wide, yellowish to greenish-yellow, and dark spotted or blotched at its base. The flower is stalked and solitary from long leaf axils, and 5 long, green sepals significantly embrace its base. It has 5 petals that are united into a widely spreading bell shape. The anthers are heavy and bluish or violet. The leaves are up to 3 1/2 inches long, alternate, and long-stalked, smooth or irregularly toothed.

Location: This plant is found, sparsely, on the north central slope of the Mound, about 100 yards from the summit.

Season: This plant has a long blooming season, from May through the fall.

Comments: The 5 sepals that are united at their base during flowering slowly grow together for their entire length, forming a five-angled, hairy, inflated "pod" or bubble around the small, ripened fruit. These delicate pods dry out, turn brown, and persist into the winter months. The Greek word *Physalis* refers to this pod.

Pink Prairie Rose
Rosa setigera

APR–JUL

Rose family (Rosaceae):
Climbing Rose; Prairie Rose; Fuzzy Rose; Wild Pink Rose

Description: This is a climbing, sprawling plant, spreading up to 15 feet, often intertwined with the undergrowth. It is covered with curved, flattened, broadbased prickles (sometimes with bristles) and densely matted hairs. The five-petaled pink flowers (fading to white) grow up to 2 inches across. They each have many crowded yellow stamens at the center. The petals are indented at their apex. There may be 5–15 pink buds or flowers in a cluster at the end of the stem, with 1 or 2 flowers open at a time. The leaves have 3–5 pointed leaflets that are up to 3 inches long, deeply wooly underneath, with deeply toothed edges.

Location: These are found in many locations all over the Mound. They are found in the northeast section, and there is a large colony on the eastern fence line. The residual red rose hips are evident until December.

Season: May is the best time to view these blooms.

Comments: When the petals fall off, the rose hips (bright red "berries") remain and are consumed by birds. Wild rose hips are high in vitamin C and were eaten by settlers and Native Americans.

Snake Herb
Dyschoriste linearis

APR–JUL

Wild Petunia family (Acanthaceae):
Narrow Leaf Snake Herb

Description: This is a rigidly upright, coarsely and stiffly hairy plant growing to 12 inches high. The stems are four-angled, usually unbranched, with several of them growing from a common base. The lavender flowers grow to 1 3/8 inches long and 1 inch across. They are hairy on the outside, two-lipped, without stems, and have dark spots or stripes in the throat. The leaves are oblong, up to 2 inches long, opposite, rather rigid, and without stems. The margins of the leaves are often fringed with hairs and the underside midribs are hairy also.

Location: There is a large colony of this flower on the northwest slope of the Mound, near the top.

Season: Flowers can continue blooming until the early fall if there is sufficient moisture.

Comments: This plant is very similar to our Wild Petunias, also in the front area. The Snake Herb tube at the base of the flower is very short while the Petunia is very long.

Poison Ivy
Toxicodendron (Rhus) radicans

APR–MAY (sometimes later)

Sumac family (Anacardiaceae):
Poison Oak; Hiedra

Photo courtesy of Jack Neal.

Description: The plant varies from not climbing to a high-climbing, very fibrous (hairy-looking) vine. Its leaves are alternate, compound, normally with 3 leaflets that are up to 4 inches long, each with a narrow pointed apex, deeply lobed, entire or sharply toothed, wavy edged, usually shiny above and sometimes hairy below, growing on a long, often purple, leaf stem. Leaves turn a conspicuously bright red in the fall. The five-petaled flowers are small (1/8 inch wide), greenish-yellowish-white, forming loose flower clusters to 3 inches long that grow from leaf axils (where the leaf joins the stem). The fruit is a smooth, white, berrylike drupe, to 1/4 inch in diameter. The fruit appears in August and persists through the winter.

Location: It can usually be found beneath any of the trees on the Mound, as the shade they create kills the prairie grasses and the ivy invades.

Season: Poison Ivy blooms early, in April and May.

Comments: "Leaves of three, let it be." Poison Ivy is a toxic plant with oil that can cause a severe allergic contact dermatitis in some individuals, anywhere from 12 hours to 5 days after exposure. It should be #1 on a list of plants to avoid. Physical contact, exposure to smoke from the burning plant, or contact with animals that have been exposed are common means of exposure. The names Poison Ivy and Poison Oak are used interchangeably and are not separate species. The sap of a relative of Poison Ivy in the *Rhus* family provides oriental lacquer.

Green Milkweed
Asclepias viridis

MAY–SEP

Milkweed family (Asclepiadaceae):
Antelope Horns

Description: This is an upright, stout, essentially smooth plant growing to 3 feet in height. Each plant has many flowers that are clustered together in one or more very distinctive, large terminal and/or lateral umbels. (An umbel is a rounded or flat-topped cluster of flowers on stems that radiate from the main stem.) Each individual flower is 5/8 inch long, pale green, with purple in its center. The leaves are over 5 inches long, broadly oval-shaped, alternate, short-stalked, very thick, firm, with smooth margins.

Location: This plant is found on the western half of the southern slope of the Mound. It is also found in the northeast corner. It is the most common Milkweed.

Season: The Milkweed is in bloom by May and continues through the summer.

Comments: The Milkweed family gets its name from the thick white sap that oozes from any broken surface. This substance can cause dermatitis in susceptible individuals. Many, if not most, of the 100 species of Milkweed are poisonous. All of them are distasteful to livestock. This plant produces two distinctive, large (4 inches long), boat-shaped pods that contain seeds tufted with silky hairs. Pioneers used these hairs to make candlewicks. This plant can be distinguished from the Green Flower Milkweed, also on the Mound, which has a more angular stem containing multiple umbels. The leaves of the Green Milkweed are less hairy and more leathery/firm than those of the Green Flower Milkweed.

Common Vetch
Vicia sativa

APR–MAY

Legume or Pea family (Fabaceae):
Spring Vetch

Description: This upright to sprawling plant has one or two pink or purple flowers growing from the leaf axils. The flower frequently appears to be folded. The stem usually terminates in branched or simple tendrils that help this plant to climb. It has alternate leaves, with each leaf blade divided into 4–8 opposite pairs of leaflets, many being blunt-tipped with tiny bristles on the tip.

Location: This plant is found in disturbed areas, especially about midpoint along the eastern fence. It can also be easily found on the west fence near Tom Thumb.

Season: The flower blooms only for a very short time, perhaps less than a week.

Comments: An invader from Europe, the family name refers to its tendrils that bind other plants.

Woods Corn Salad
Valerianella woodsiana

APR–MAY

Valerian family (Verbenaceae)

Description: This plant grows to about 20 inches in height. It has a solitary stem with opposite, forking, almost symmetrical branches near the upper portion of the stem. There are numerous tiny, five-lobed, funnel-shaped flowers crowded together in rounded terminal spikes on each branch. Each flower has 3 stamens that protrude significantly. The leaves grow up to 2 1/2 inches in length. The leaves are simple, long, narrow, opposite, and clasping, with hairy margins. The upper leaves are sometimes toothed near their base.

Location: This plant is found near the west fence at about its midpoint.

Season: Corn Salad starts blooming in April and continues through May.

Comments: Some species of the Valerian family have been used medicinally to treat nervous disorders. This is a very tiny flower that one might miss at first glance.

White Avens
Geum canadense

APR–MAY

Rose family (Rosaceae):

Description: This plant grows to 2 1/2 feet tall, with multiple hairy stems growing from the base. The stem hairs lie flat against the stem, especially near the base of the plant. The solitary flowers have 5 small (to 3/4 inch across) widely spaced, white petals with 5 very distinctive, slightly shorter, green sepals. The leaves are alternate, with 3 large terminal leaflets and 2–4 much smaller ones near the base. The leaflets are sharply toothed and occasionally lobed near the base of the stem.

Location: Look under the large ash trees along the east fence, about halfway down.

Season: White Avens blooms early in April and May.

Comments: The three-part leaf is typical of the Rosaceae (rose) family. The fruit of this plant has large hooked bristles that cling to whatever touches it.

Black Eyed Susan (Brown Eyed Susan)
Rudbeckia hirta

MAY–NOV

Sunflower family (Asteraceae):
Brown Eyed Susan

Description: This is a stiffly upright, coarse, extremely bristly, hairy plant occasionally forming a clump and growing to 3 feet in height. Our plants grow to about 8 inches in height. The flower head is up to 3 inches across, solitary, and terminal on a long, slender stem. The golden ray flowers (8–20) are solid yellow or with brown in the basal portion and are wide-spreading. The disk flowers are velvety, chocolate-brown, numerous, forming a small, flattish cone. The leaves grow up to 7 inches long and up to 3 inches wide, are rough, alternate, pointed and quite hairy, with short stalks. The lower leaf blades are entirely to somewhat toothed with three prominent veins.

Location: This plant is found primarily in the northeast corner of the Mound.

Season: Black Eyed Susans have a long blooming period, from May to November.

Comments: The Cherokee Indians used the roots to treat earache. Some people make tea from the dried flowers and leaves. When used in dyeing, it produces greenish and yellowish colors. This plant has been reported poisonous to grazing animals.

Lance Leaf Loosestrife
Lythrum alatum (lanceolantum)

MAY–JUL

Looseleaf family (Lythraceae):
Purple Loosestrife; Winged Loosestrife

Description: This plant grows to 5 feet or more in height. Its stem is slender, brittle, and four-angled, usually branching in the upper portion. The plant has numerous creeping basal offshoots that eventually form large colonies. The beautiful lavender-blue to reddish flower grows up to 1 inch across. It has 4–6 petals that are wrinkled (like crepe paper) and often paired, spreading flat. They grow in crowded, terminal, spike-like clusters on the upper third of the stem. The leaves are up to 2 1/2 inches long and about 1/2 inch wide. They are simple, often whorled and folded against the stem.

Location: There are two large colonies of this prominent plant on the southern slope, near the east gate, about thirty yards from the fence line.

Season: Loosestrife is found blooming from May to July.

Comments: This plant is frequently used in Mexico as an astringent.

Rain Lily
Cooperia drummondii

MAY–SEP

Lily family (Liliaceae):
Cebolleta; Evening Star; Drummond Rain Lily

Description: The plant is upright, growing up to 9 inches in height from a deep bulb. The flower is white (sometimes with a pink tinge), fragrant, solitary, growing at the end of a fragile, leafless stem. It has 3 petals and 3 sepals that are the same color as the petals. The flower is trumpet-shaped after first opening, but soon it spreads widely to about 2 inches across when mature. The basal leaves elongate after flowering and become very narrow, almost grasslike.

Location: After significant rainfalls, large numbers of this plant have been seen in the disturbed areas along the eastern and northern fence line.

Season: The flowers appear irregularly from late spring to early fall.

Comments: The flowers appear two or three days after a rain. They begin to open in late afternoon or evening and are fully open the next morning. They last up to 4 days before turning pinkish and withering.

Common Sunflower
Helianthus annuus

MAY–NOV

Sunflower family (Asteraceae):
Mirasol

Description: This is a prominent, upright (to 8 feet tall, but exhibiting much variation in height), stout, very rough, hairy plant. Its solitary stem is bristly and sticky to the touch. Although quite variable in size, it flowers even when very short. The flower head grows up to 5 inches across. There is a solitary bloom on each plant that is terminal on a long, rough, hairy stalk. It has 20–30 infertile, deep yellow, ray flowers that are up to 2 inches long and often overlapping. The disk flower is brown-red and up to 2 inches across. There can be as many as 30 flower heads on each plant, each at the tip of a separate stem. The leaf blades are up to 12 inches long, dull green, long-stalked, rough to the touch, margin-toothed, broad at the base and tapering to a point. They are mostly alternate, except that the lowermost leaves are often opposite.

Location: The Common Sunflower is found virtually all over the Mound.

Season: The plant can be seen in bloom for most of the summer and fall. It is one of the last plants to die.

Comments: The flower heads "follow the sun," facing east in the morning and west in the evening. One of the first cultivated plants in North America, it is still grown commercially for its sunflower seeds, an important source of oil and food products. Native Americans used it for medicine and dye, as a source for fiber, and as a nutritious food. It can be easily distinguished from our other sunflower, the Maximilian Sunflower, because it has only one flower per stalk. The Common Sunflower is the state flower of Kansas.

Field Pansy
Viola rafinesquii or V. bicolor

FEB–APR

Violet family (Violaceae):
Wild Pansy; Johnny-Jump-Up; Heart's Ease; Cupid's Delight

Description: This is a rather low plant (up to 1 foot in height) with a solitary stem and very fine hairs. The flowers (less than 1 inch across) grow on long stems. They have 5 unlobed petals that vary in color: violet, blue, lavender, or white. The upper 2 petals are large and erect; the lower petals are paler and darkly veined toward the base; the lowest petal has some yellow in the throat of the flower. The alternate leaves grow to 1 3/4 inches long, in clusters on the stem. The basal leaves are long-stalked and round. The leaves on the upper stem are smaller and narrower, with smooth edges.

Location: Only one plant has been documented; it was in the north central section and is reproduced here. This is the most compelling reason for not picking flowers on the Mound.

Season: Uncertain.

Comments: In other areas, this plant is known to produce large colonies. Parts of the plant were used medicinally.

Prairie Acacia
Acacia angustissima

MAY–OCT

Legume, Pea, or Bean family (Fabaceae):
Fern Acacia; White Ball Acacia

Description: This is a branching, trailing plant bearing round balls of white flowers on long stalks arising on the axils of compound leaves. It sometimes forms small clusters. The individual flowers themselves are quite tiny, crowding in very dense, white, round clusters about 1 inch across at the tip of the stalk. The leaves are alternate, up to 3 inches long, divided into 4–12 pairs of small segments, with these again divided into 6–20 pairs of tiny leaflets (fernlike—hence "Fern Acacia"). The leafstalks themselves are prickly, which distinguishes this plant from other similar ones. Each plant forms one or more classic bean pods, each containing 3–8 seeds. The pods change color from green to reddish-brown as the season progresses.

Location: This plant is found in many small colonies, mainly on the southern slope of the Mound.

Season: Spring or later, depending on rains.

Comments: The Prairie Acacia is high in protein and is readily eaten by livestock. The leaves are sensitive and will fold when touched. The flower is round like the Sensitive Brier except that the Sensitive Brier is pink in color.

White Prairie Rose
Rosa foliolosa

MAY–JUL

Rose family (Rosaceae):
Leafy Rose

Description: This is a sprawling plant, often intertwined with the undergrowth. It is covered with relatively few, slender prickles. The white (rarely light pink) flowers are solitary on a short stem. Each flower has 5 very delicate petals that are up to 2 inches across, with numerous yellow stamens crowded in its center. The leaves have 3–5 pointed leaflets, up to 3 inches long, and deeply toothed.

Location: This flower is found mainly near the north end and halfway down the east side of the Mound.

Season: Blooms through the summer.

Comments: When the petals fall off, you will see bright red rose hips that persist into the winter, one to a stem. If you see a number on one trailing stem, it was the pink version.

Clasping Leaved Coneflower
Dracopis amplexicaulis

MAY–JUN

Sunflower family (Asteraceae):
Coneflower

Description: This plant grows up to 2 1/2 feet in height. It has a smooth, solitary stem, branching in its upper portion. It is covered with a thin, whitish coating that comes off when rubbed. The solitary flower head is found at the end of a long, slender branch and is up to 2 inches across. It has 5–9 ray flowers (petals) that are up to 1 inch long and yellow or, more commonly, yellow with a reddish-brown basal portion. The erect, elongated, brownish central disk flower is conical or oblong and up to 1 1/4 inches tall. The opposite, stalkless (or nearly so) leaves grow up to 4 inches in length and occur mostly on the lower portion of the plant where they tend to clasp the stem. They are not lobed and are relatively stiff.

Location: This plant is found all over the Mound, particularly on the southern slope.

Season: Best show for this flower is May through June.

Comments: The flower of this plant looks similar to that of the Mexican Hat except the leaves. The leaves of this plant tend to wrap around or "clasp" the main stem, to which its Latin name refers. Native Americans were able to extract a red dye from the plant for dyeing woolen yarn.

Chapter 6 Explosion of Color

Bindweed
Convolvulus equitans

MAY–JUL

Morning Glory family (Convolvulaceae):
Texas Bindweed; Gray Bindweed

Description: This is a low, climbing, vine-like plant that can form extensive beds. Its stems grow to 6 feet long, not rooting where they contact the ground. The flowers are up to 1 1/8 inches long and 1 inch across. They have 5 white to pale pink petals that are united at the base to form a broadly flaring trumpet shape. The center is pink to red, with 5 yellow stamens and 1 yellow pistil. Flowers grow from leaf axils on a long stem. The leaves grow to 2 3/4 inches long and 1 1/2 inches wide. The leaf blades are variously shaped (toothed to lobed to entire), with dense silky-wooly gray hairs on both surfaces. They are mostly palm-shaped, with their margin having 5 sharp points or projections.

Location: This climbing plant is found all over the Mound, especially on the southern slope near the crest. The variations are mostly at the front fence.

Season: A warm weather plant, it blooms all summer from May to July.

Common Bindweed

Comments: This plant is a bane to farmers. Especially problematic in Kansas wheat fields, it is able to strangle the stems of cultivated plants and sometimes grows in such profusion as to cut off the sunlight, causing the cultivated plants to die. It is very difficult to eradicate because of its deep root system. *Convolvulus* refers to the Latin, "to entwine."

Variations: A second Bindweed on the Mound is *Convolvulus arvensis*, a non-native invader (top right). A third is Hedge Bindweed, *Calystegia sepium*, a wild Morning Glory (bottom right).

83

Basket Flower
Centaurea americana

MAY–JUL

Sunflower family (Asteraceae):
American Basket Flower; Thornless Thistle; Shaving Brush; Star Thistle

Description: This plant grows up to 5 feet in height. The stem is solitary, grooved or ridged, thick, leafy, and has many branches toward the top. The large, dark lavender to pink, fragrant, solitary flower head (up to 3 inches across) has no ray flowers (petals). The outer borders of the thread-like disk flowers are somewhat longer and darker in color than the inner ones, and they tend to droop. The flower head appears to sit in a basket or cup-like structure consisting of green, prickly bracts. The stem is enlarged or swollen beneath the bracts and the flower head. The leaves grow up to 2 1/2 inches long, are lance-like, alternate, stemless, with smooth or shallowly toothed edges, with no spines.

Location: This plant is found in great numbers all over the Mound.

Season: This plant has the same bloom season as the Texas Thistle, so you can readily compare for positive identification.

Comments: This plant looks much like a classic thistle, but it lacks the prickly characteristics of the thistle. Before it is fully open, the flower looks somewhat like a classic shaving brush. The dried flower head remains on the stalk for many months, making this plant easy to identify once the colors of the flower have gone. It is used as a dried ornamental.

Green Flower Milkweed
Asclepias viridiflora

JUN–AUG

Milkweed family (Asclepiadaceae):
Green Antelope Horns; Wand Milkweed

Description: This is an upright, stout, relatively hairy plant growing to 3 feet in height. The stem is quite angular. Each plant has many flowers that are tightly clustered together in one or more very distinctive, large terminal and/or lateral umbels. (An umbel is a rounded or flat-topped cluster of flowers on stems that radiate from the main stem.) Each individual flower is 5/8 inch long, round, pale green, with purple in its center. The leaves are quite variable, over 5 inches long, pale green, hairy, alternate, and short-stalked, with wavy, hairy margins.

Location: This plant is found on the western half of the southern slope of the Mound. It is also found in the northeast corner.

Season: Green Flower Milkweed flowers later, in June through August.

Comments: The Milkweed family gets its name from the thick white sap that oozes from any broken surface. This substance can cause dermatitis in susceptible individuals. Many, if not most, of the 100 species of Milkweed are poisonous. All of them are distasteful to livestock. This plant produces two distinctive, large (4 inches long), boat-shaped pods that contain seeds tufted with silky hairs. These hairs were formerly used to make candlewicks. This plant can be distinguished from the Green Milkweed by its angular stem with multiple umbels. The leaves of this plant are more hairy and less leathery/firm than those of the Green Milkweed.

Indian Blanket
Gaillardia pulchella

MAY–JUN (sporadically later)

Sunflower or Daisy family (Asteraceae):
Firewheel; Rose-Ring Gaillardia; Blanket Flower

Description: This is an upright, sprawling, hairy plant that grows up to 2 feet in height. The flower heads can grow larger than 2 inches across and are solitary at the end of long, slender stalks. Each flower has 10–20 ray flowers (petals), reddish (occasionally all yellow) in color with a narrow to wide orange-yellow band at the tip. The central disk flowers are numerous, purplish-red to brown. The pointed leaves can exceed 3 inches in length. They are alternate, with short stalks. They tend to clasp the stem at the base of the flower and are variously entire, toothed, or deeply lobed.

Location: This prominent plant grows all over the Mound, but mainly on the southeast slope.

Season: This plant has a long blooming season. Some individuals can be seen as early as mid-April, but in May they are in large colonies. Isolated specimens can be found well into the fall.

Comments: This is the state wildflower of Oklahoma. The seeds are readily available because of its popularity. It is a native only of the tall grass prairie in Oklahoma and Texas.

Texas Thistle
Cirsium texanum

MAY–JUL

Sunflower family (Asteraceae):
Southern Thistle

Description: This is a tall (up to 5 feet) bristly-spiny, woolly plant with a solitary stem that branches toward the top. The pink or rose-purple flower head grows up to 2 1/2 inches across at the tip of a long, woolly, almost leafless stalk. There are no ray flowers (petals). The flower head, consisting of many hundred tiny disk flowers, appears as though it is held in a 3/4-inch high cup of small, prickly tipped bracts. The leaves are long (up to 9 inches), alternate, numerous, stalkless, and clasp the stem at their base. They are green, smooth on top but covered on the underside with whitish, felt hairs. The leaves have several irregular lobes, with a pointed tip and prickers.

Location: This plant is found all over the Mound, particularly along the east fence line and on the southern slope.

Season: Emerging in May and continuing on into July, this plant is plentiful and easy to identify.

Comments: The flower appears to be similar to that of the Basket Flower, but this plant has leaves with sharp needle-like points and spines. Its botanical name refers to its medicinal use for swollen veins.

Color Chart
April–May

White

- Yarrow
- Yucca
- Prairie Larkspur
- Indian Plantain
- Old Plainsman
- Green Milkweed
- Poison Ivy
- Woods Corn Salad
- White Avens
- Rain Lily
- White Prairie Rose
- Prairie Acacia
- Bindweed
- Green Flower Milkweed

Yellow

- Prairie Parsley
- Yellow Sweet Clover
- Engelmann Daisy
- Goatsbeard
- Yellow Ground Cherry
- Black Eyed Susan
- Common Sunflower
- Clasping Leaved Coneflower

Blue

- Locoweed
- Ohio Spiderwort
- Texas Vervain
- Venus' Looking-glass
- Western Horsenettle
- Lemon Beebalm
- Blue Sage
- Snake Herb
- Field Pansy
- Basket Flower
- Texas Thistle

Red

- Wild Garlic
- Prairie Phlox
- Wild Foxglove
- Indian Paintbrush
- Sensitive Brier
- Scarlet Gaura
- Pink Prairie Rose
- Common Vetch
- Lance Leaf Loosestrife
- Firewheel

87

Chapter 7
Drought *Tolerant*
The Flowers of June through November

This period on the Mound is only for the adventurous, particularly in July and August. It is the time of the blazing Texas sun when temperatures are likely to be one hundred degrees or more and there is no shade on the prairie. Wide-brim hats and sunscreen are recommended. Still, it's likely to be a short trip. Aside from the heat, most plants have reached their maturity, so walking is difficult. The grasses and brambles will slice and scratch bare skin; long pants and boots are really necessary. If that's not enough, the insects are abundant. There are numerous wasps and bees, and the fire ants are especially vindictive in the heat. Though seen only once, there is a copperhead on the species list. But this is only a warning, as the Mound is still covered with many diverse and colorful plants that are perfectly happy with these harsh conditions. To the adventurous go the rewards of finding the following group of plants.

Indian Blanket

Prickly Lettuce
Lactuca serriola

JUN–AUG

Sunflower family (Asteraceae):
Compass Plant; Wild Lettuce

Description: This is a tall plant (to 5 feet) with a white, erect stem. The foliage exudes a milky sap when broken. There are loose clusters of numerous, small, yellow, dandelion-like flower heads on each plant. Each flower head is up to 1/2 inch across, usually with 16–24 petals (ray flowers). The leaves are alternate, relatively large (2–12 inches long), oblong, coarsely toothed, lobed or unlobed, and prickly edged, with numerous yellow prickles on the underside midrib of most leaves. The bases of the leaves clasp the stem. A very distinctive trait of this plant is that the leaves are often oriented together in an approximate, flat, north-south plane.

Location: This plant is found, singly and in very small groups, all over the Mound, particularly near the borders.

Season: Wild Lettuce is a hot weather bloomer, from June through August.

Comments: The small but beautiful yellow flower on this plant is in bloom only for a very few hours each morning, ordinarily between 8:00 a.m. and 11:00 a.m. If you miss it, you will either find numerous closed buds or a large number of fluffy dandelion-like seed head clusters on each plant. Cattle have been poisoned by eating large quantities of this plant. This wild lettuce is seen all over town. *Lactuca sativa* is common lettuce and doesn't have the bitter taste of wild lettuce.

Musk Thistle
Carduus nutans

MAY–SEP

Sunflower, Aster, or Daisy family (Asteraceae):
Nodding Thistle

Description: The plant itself grows up to 3 feet tall. The thistle-like, rose-purple flower head is up to 2 1/2 inches across and grows singly at the end of a long floral stalk at the end of the main stem. A cobweb-like covering may be evident on the stem. Particularly in the early stages of growth, broad, pointed, purple bracts surround the base of the flower head, with the outer ones curving outward. The leaves grow up to 10 inches long and are very spiny. The bases of the leaves extend up and down the stem, resembling prickly "wings."

Location: This plant may be scattered throughout the Mound. There are usually a number of plants at the west gate.

Season: Plants are blooming in early May.

Comments: Everything about this plant is "thorny." It is a European invader and is very aggressive.

Wood Sage
Teucrium canadense

JUN–AUG

Mint family (Lamiaceae):
American Germander

Description: This is an upright plant growing to 3 feet in height, although those on the Mound appear to grow only to about 1 1/2 feet in height. It has a solitary, hairy stem that is prominently four-angled and frequently branched in its upper portion. The flower grows to 3/4 inch long and is fragrant and pale pink or lavender, with purple spots. The lower lobe of the flower is long and flattened; its upper lobes are short. The flowers on each plant are numerous and congested toward the top in dense, terminal, elongated spikes. The leaves grow to 4 inches long and 1 1/2 inches wide, are opposite, lanceolate, and stalked, with toothed margins. They are hairy on their lower surface.

Location: This plant is located in large numbers toward the northeast corner of the Mound, approximately 50 yards from the north fence and also midway to the fence on the north central slope.

Season: Wood Sage prefers to bloom in the hottest period, June through August.

Comments: This plant usually forms dense colonies due to creeping rhizomes.

Wild Petunia
Ruellia humilis

JUN–SEP

Wild Petunia family (Acanthaceae):
Low Ruellia; Fringe Leaf Ruellia; Zigzag Ruel; Prairie Petunia

Description: This is an upright, stout, hairy, somewhat sprawling plant growing to 32 inches in height, often growing in clumps. Its stems are prominently four-angled. Its branches are slender and often found lying on the ground. Five sepals that are united at the base and constitute up to 1/3 of the flower's length surround the flower. The lavender to bluish trumpet-shaped flower is up to 3 1/8 inches across. It has 5 deeply lobed petals and often has reddish or purple spots down one side of the throat. Each flower opens in the early morning and lasts only one day. There are few flowers on each stem, growing from the middle and upper leaf axils. The leaf blades are lance-shaped, leathery, and opposite. They have little to no stalk and grow upright close to the stem. They are densely hairy on their veins and margins (hence "Fringe Leaf Ruellia").

Location: There is a conspicuous group of this beautiful plant on the eastern slope of the Mound, about halfway to the east fence.

Season: The Wild Petunia is a hot weather plant, blooming June through September.

Comments: Refer to Snake Herb if difficult to distinguish. In the afternoon, there are usually fallen blooms present as they only last a day.

White Prairie Clover
Dalea multiflora

JUN–SEP

Legume, Bean, or Pulse family (Fabaceae):
Round Head Dalea; Round Head Prairie Clover

Description: This is a low, sprawling, bushy plant with many branches, growing up to 3 feet tall. Its tiny white flowers are up to 1/4 inch long, with 5 petals and 5 stamens. They are grouped into a dense, small, cylindrical, terminal spike up to 3 inches tall and 1/2 inch in diameter, somewhat similar in shape to the flower of clover one finds in one's yard. The leaves are up to 4 inches long, on a short stalk, with 3–9 small, hairless, bright green leaflets (up to 1/2 inch long) on each blade.

Location: This plant is seen in numerous locations on the Mound, to include sporadically along the eastern, northern, and southern fence lines.

Season: White Prairie Clover is in full bloom by the middle of June.

Comments: The bright white flower balls disappear and leave a hairy, bright red fruit.

Common Broomweed
Gutierrezia dracunculoides

JUL–NOV

Sunflower family (Asteraceae)

Description: This is an upright plant growing to 3 feet high. It has a solitary stem that is often sticky to the touch and that is densely and intricately branched in the upper portion. The plant grows in a mound-like clump of crowded, yellow flowers. Each small flower head is solitary at the end of a short branch and grows to 5/8 inch across. It has 7–15 oblong or oval shaped (petal-like) yellow ray flowers. Its central disk flower is also yellow. The leaves are alternate, stalkless, very narrow, up to 2 inches long, dotted with small glands, and pointed at the tip. By the time the plant blooms, many of the leaves have fallen off.

Location: There are a few Common Broomweed plants on the southern slope near the crest of the Mound.

Season: Fall into early winter. This is one of the last yellow flowers evident in the growing season.

Comments: Dried Common Broomweed plants were tied together by early settlers and used as brooms (hence the name). This plant can cause contact dermatitis in humans.

Maximilian Sunflower
Helianthus maximiliani

JUL–OCT

Sunflower family (Asteraceae):
Michaelmas Daisy

Description: This is an upright, tall (to 10 feet), leafy, rough-hairy plant that often grows in clumps. It usually has several, mostly unbranched, stems growing from a common base. There are few to numerous flower heads growing on the upper portion of the plant in a spike or tower-like arrangement. Each flower grows up to 1 inches across at the end of a short stalk. The 15–19 petal-like ray flowers are yellow and pointed. The disk flowers are a darker yellow and are up to 1 inch across. The leaves are to 12 inches long (shorter toward the top), 2 inches wide, alternate, and are often conspicuously folded lengthwise and curving downward. They are tapered at both ends, grayish-green, stiff, rough to the touch on both sides, and sometimes slightly toothed.

Location: There are several specimens growing on the southern slope of the Mound, near the fence.

Season: This seems to be one of the last flowers to bloom, and it is one of the last to fade.

Comments: Livestock readily graze this plant. Deer and birds eat the seeds. The multiflowered stem easily distinguishes the Maximilian Sunflower from the Common Sunflower. It is named after Prince Maximilian. It is one of the prettiest and most common sunflowers.

Western Ironweed
Vernonia baldwinii

JUL–SEP

Sunflower family (Asteraceae):
Baldwin's Ironweed

Description: This is an upright, stout, coarse, hairy plant that can grow to 5 feet tall. The plant is spread by underground runners and tends to grow in clumps. Each individual flower grows up to 1/2 inch high, with perhaps 40 of them grouped together into a round, purple flower head less than 1 inch across. There are no ray flowers (petals). These purple flower heads in turn are grouped into a large, stalked, compact, terminal cluster. Several such clusters can be found on the same plant. The 6-inch long leaves are distributed along the entire stem. They are stalkless, alternate, lance-shaped, deeply toothed, with hairs on the lower leaf surface.

Location: There is a colony of Western Ironweed plants on the south-central and southeast slopes, close to the fence.

Season: Ironweed emerges as most other plants are dying off and flourishes from July through September.

Comments: This is one of our most robust and drought-resistant plants. Its bitter-tasting foliage keeps it from being eaten by cattle. It was named for its discoverer, William Baldwin, a Pennsylvania botanist. Like the Gayfeather, it provides dramatic late fall color and is easy to identify.

Saw-leaf Daisy
Grindelia papposa

JUL–OCT

Sunflower family (Asteraceae):
Goldenweed

Description: This plant is stout, stiffly upright, tall (to 5 feet), often sticky to the touch, and usually unbranched except near the top where the flowering portions are. The flowers are about 1 1/2 inches across with numerous yellow petal-like ray flowers. The central disk flowers are also yellow. The flower heads themselves are few to very numerous, short-stalked, and very congested near the top of the stalk. Its leaves are up to 3 inches long, oblong, alternate, stalkless, clasping the stem at the leaf base, thick, stiff, and, most distinctively, very coarsely spine-toothed.

Location: This plant is found on the southern slope of the Mound, primarily toward the front fence and at the top, especially on the western side.

Season: This plant appears in the hot weather in July and continues to October.

Comments: This drought-resistant plant is not grazed by livestock and usually forms large, conspicuous stands.

Wild Four O'Clock
Mirabilis linearis

JUL–OCT

Four O'Clock family (Nyctaginaceae):
Linear Leaf Four O'Clock

Description: This is a tall (up to 4 feet), erect, spindly plant with very few leaves on the upper portion of the slender, quite angular, stem. The flowers consist of 5 deeply grooved, joined, funnel-shaped, pink-purple, petal-like sepals growing up to 1 inch in length. Each flower has 5 protruding, prominent pink or purple stamens with bright yellow anthers. The flowers themselves are very fragile, opening in the late afternoon and dropping off early the next morning. Flowers and branches grow out of the leaf axils (where the leaf joins the stem). The leaves are opposite and much longer than wide.

Location: This plant has been seen in many locations on the Mound: near the south fence, in the center; on the crest; and in the northwest corner near its fence.

Season: This flower starts blooming in the heat of the summer and continues till late in the fall.

Comments: Flowers of the Four O'Clock family lack petals but have brightly colored sepals that may be mistaken for petals. Once the colorful sepals drop off, the remnant becomes a prominent, star-shaped growth, becoming translucent, "papery," and rose-colored with age, as the fruit seed matures. This growth is frequently mistaken for a flower. Whether petals or sepals, it is a beautiful display and well worth the search.

Purple Gerardia
Agalinis purpurea

AUG–OCT

Foxglove, Figwort, or Snapdragon family (Scrophulariaceae)

Description: This plant is very finely stemmed, sprawling (almost vine-like), much branched toward the top, growing to 4 feet or more. The delicate, funnel-shaped flower blooms from the leaf axils. It is 1 1/8 inches long, purplish to light pink, with 5 fused, unequal flaring lobes at the rim. The flower has a short stalk, 2 yellow lines, small, darker purple spots in the throat, and a conspicuously bearded base. Its narrow leaves are up to 1 1/2 inches long, opposite, and stalkless.

Location: This is a very low, crawling plant with a small flower that might be hidden in the undergrowth. Specimens have been found near the south fence and on the summit of the Mound.

Season: Gerardia blooms late, from August to October.

Comments: Native Americans created a medicinal potion from the roots of this plant. There are several species of this plant that differ only slightly.

Heath Aster
Aster ericoides

SEP–NOV

Sunflower family (Asteraceae):
White Prairie Aster; Wreath Aster

Description: This plant is upright to reclining, with many, many branches (almost shrub-like in large colonies that are formed by underground runners). The short-stalked flower heads are symmetrical, up to 1/2 inch across, and are often very crowded on a branch. The 15 or so (petal-like) fertile ray flowers are normally white; the central disk flowers are yellow to rose. The larger basal leaves have often withered away by the time the plant blooms. The remaining leaves on the flowering branches are small, alternate, numerous, and stiff.

Location: It is found almost everywhere on the Mound in its time. There is an especially large colony near the front fence, toward the east side.

Season: This is one of the latest blooming plants on the Mound, lasting until the first frost.

Comments: This flower differs from the somewhat similar Prairie Fleabane in that the ray flowers on the Heath Aster are not so numerous and are larger. It inhabits a wide variety of prairie systems. There are many Native American medicinal uses for the asters.

Ragweed
Ambrosia trifida

AUG–NOV

Sunflower family (Asteraceae):
Great Ragweed; Giant Ragweed; Buffaloweed; Blood Ragweed

Description: This is a very tall (to 10 feet or more), upright, coarse, hairy-stemmed plant with inconspicuous flowers growing in multiple elongated clusters on the same stem. The sap is blood red. The plant is unisexual, with minute male and female flowers produced on the same plant. Male flowers are tiny, yellow-green, packed with yellow stamens, borne in heads of 15–20 florets each, arranged in elongated (to 10 inches long), slender clusters. The female flowers are small, green, and stalkless and are found in small clusters at leaf axils. The leaves are large (up to 8 inches long), opposite, palmately three-lobed, and rough to the touch on both sides.

Location: This plant is found virtually all over the Mound. It is one of our tallest plants, and it is one of the last to die.

Season: Ragweed is a late bloomer. Hay fever sufferers know when this plant blooms, from August to November.

Comments: This plant is a problematic cause of allergic reactions during the fall and is considered to be the leading cause of hay fever. The plant is wind-pollinated, with huge quantities of pollen grains released normally in midmorning, as the dew dries and the humidity decreases. Native Americans have used ragweed as medicinal tea.

Goldenrod
Solidago altissima
Solidago canadensis

AUG–NOV

Sunflower family (Asteraceae)

Description: This is a tall, stiffly upright, coarse, hairy plant growing to 7 feet or more. It grows in large colonies via underground roots. Each yellow flower is tiny (1/8 inch high), but the flower heads contain many hundreds of flowers in large, loose, plume-like or pyramidal terminal clusters as well as many short arching branches. The leaves are up to 6 inches long. They are alternate, stalkless, a dull-green, rough above, hairy below, with toothed margins toward their tip.

Location: This plant is prevalent all over the Mound. Late in the season, after other plants have gone dormant, this plant is still prominent everywhere. It is one of the last to go dormant.

Season: The blooming season is from late summer into late fall. This is an Indian Summer plant.

Comments: It is extremely difficult to distinguish among many of the 80 species of *Solidago* Goldenrod. These plants are insect-pollinated and thus are not particularly significant in emitting windblown pollen, which causes asthma or hay fever. Most such allergies are likely attributable to Ragweed, which flowers at the same time as Goldenrod. Medicinal uses of Goldenrod are many; it has been used to make tea for digestive ailments as well as for making natural yellow dye.

Gayfeather
Liatris mucronata

AUG–NOV

Sunflower family (Asteraceae):
Narrow Leaf Gayfeather; Blazing Star;
Button-Snakeroot

Description: This is a distinctive, stiffly upright, unbranched, purple flowering plant that is up to 32 inches tall. Each individual purple flower is tiny (1/8 inch tall). There are no petals (ray flowers). The individual flower heads are densely congested on a long, slender, terminal, upright spike that is up to 18 inches in length. The plant has very narrow leaves to 4 inches long. They are alternate, have smooth edges, are pointed and somewhat crowded, becoming progressively smaller from bottom to top.

Location: There is a cluster of these plants, covering perhaps ten square yards, on the summit of the Mound, toward the western side. Near the front fence in the center is another colony.

Season: Gayfeather opens in August, but the most perfect bloom is in September.

Comments: The bulb of this plant has produced products to treat sore throat and rattlesnake bite. The *Liatris* has had many medicinal uses: as a diuretic, a mild kidney or liver tonic, and a treatment for laryngitis. It is a dramatic plant in the fall, producing a strong color.

Sow Thistle
Sonchus asper

JUN–OCT

Sunflower family (Asteraceae):
Prickly Sow Thistle; Spiny-Leafed Sow Thistle

Description: This weedy plant grows to 5 feet in height. It has a smooth, stout, hollow, angular (usually solitary) stem, often branched in its upper portion. It exudes a milky "sap" when bruised. The yellow, dandelion-like heads grow to 1 inch across. It has numerous petal-like ray flowers with no central disk flower. There is one to several flower heads on short, leafless stalks, mostly near the top of the stem, appearing clustered. The stiff leaves grow to 8 inches long and are usually stalkless with margins that are lobed and sharply toothed with spines. (Upper leaves are smaller and less lobed.) They usually clasp the stem.

Location: This plant is found primarily on the southern slope.

Season: Sporadically all year, depending on rainfall patterns and quantities.

Comments: The milky sap of this plant was used for eye ailments and skin problems. It was listed as a medicinal in Europe where it originated. It is considered a noxious weed in the United States.

Color Chart
June–November

White

- White Prairie Clover
- Heath Aster

Yellow

- Compass Plant
- Common Broomweed
- Maximilian Sunflower
- Saw Leaf Daisy
- Goldenrod
- Sow Thistle

Blue

- Musk Thistle
- Wood Sage
- Wild Petunia
- Western Ironweed
- Purple Gerardia
- Gayfeather

Red

- Wild Four O'Clock

Green

- Ragweed

Chapter 8: The Majestic Grasses

The best time to study the grasses is in the fall, the first week in October to be precise. It is then that most of the grasses are in full bloom (they have flowers, too, although very minute). The Big Bluestem, Indian Grass, and Small Bluestem dominate the prairie. Most have lost their blue-green coloring and are becoming a rich golden brown or russet, especially in the late afternoon sun or after a fall thunderstorm. Fall's temperature is pleasant, and many of the bugs are gone. The grasses are often ignored in field guides, but they are incredible, ancient plants, and understanding the coexistence of the grasses and the flowers makes a more complete prairie experience.

Big Bluestem
Andropogon gerardii

Grass family (Poaceae)

Description: Big Bluestem grows in clumps on the Mound and may be from 3 to 6 feet tall, depending on location and season. The prominent seed head is usually branched in threes, giving rise to the pioneer common name, "Turkey Foot." It is a bright blue-green color in the spring and changes to a rich russet color in the fall, especially after a rain. The long, abundant leaves are 1/2 inch wide and have minute hairs on the edges. There is a small collar at the base of the leaf at the stem.

Location: Big Bluestem can be found all over the Mound, but the best plants are near the center of the property.

Season: The grass is emerging and greening up the Mound by April. It increases all summer and flowers in late September and early October. The Prairie Picnics were always held the first weekend in October to celebrate the flowering of this dramatic plant.

Comments: This is one of the four major prairie grasses and produces excellent and abundant forage. It is drought-resistant and herbivore-resistant because its root system is up to 12 feet deep. An individual clump may be seventy-five years old. You can almost detect Bluestem blindfolded because the thick leaves make walking difficult.

Little Bluestem

Schizachyrium scoparium

Grass Family (Poaceae)

Description: Usually the dominant prairie grass in our area, Little Bluestem grows in clumps similar to Big Bluestem but is shorter overall, 2–5 feet. The leaves are smaller. It is blue-green in the spring and russet color in the fall. The flowers are many and grow along the stalks, becoming fluffy in full bloom.

Location: Distributed over the Mound, prominent near the top.

Season: Early spring (April) to late fall, until the first freeze (November).

Comments: After seeing this plant on the Mound, it is easy to identify remnants scattered about town on roadsides and fields, especially around Grapevine Lake. These prairie grasses far outproduce Bermuda grass in quantity and quality.

Indian Grass
Sorghastrum nutans

Grass family (Poaceae)

Description: This beautiful grass is one of the four major prairie grasses. It can be 5–7 feet tall on the Mound and has long, narrow leaves 1/2 inch wide. It can be separate or in small groups. The prominent, yellow, fluffy seed heads can be identified at quite a distance.

Location: Scattered throughout the Mound, with the best population at the upper part.

Season: The best time of year to identify this plant is at full bloom, mid-September to mid-October. All of the grasses are flowering and abundant then. It is worth a special visit at this time to experience the grass towering over your head and imagine when all Denton County was covered in prairie.

Comments: Indian grass is the state grass of Oklahoma. It is very desirable for livestock and indicates a healthy prairie.

Sideoats Grama
Bouteloua curtipendula

Grass family (Poaceae)

Description: Sideoats is the descriptive term of this native grass. The 25 or more seed heads will often be arranged evenly spaced along one side of the stem. Some can be found alternating on stems also, as the forms are variable. The grass is not very tall, less than 3 feet, so look for colonies close to the ground, beneath the tall plants. The leaves are about a foot long and only 1/4 inch wide. The leaves have sharp points. If you can catch it flowering, it has bright red-orange stamens.

Location: Found throughout the Mound, most populous near the top with the tall grasses.

Season: The grass is growing on the Mound by April and flowering in September and October.

Comments: Another of the four major grasses that describe a native prairie, Sideoats Grama was chosen as the Texas state grass in 1971. It is nutritious, productive, and favored by grazing animals.

Texas Wintergrass
Nassella leucotricha

Grass Family (Poaceae):
Spear Grass; Texas Needle Grass

Description: A low, clumpy fine grass, its flowers appear at the ends of long, fine stalks, about 1 foot in height.

Location: Scattered about the Mound but most conspicuous near the paths where other grasses are beaten back.

Season: Emerging in April and fully formed in October.

Comments: Considered a minor participant in the native prairie collection, it increases in disturbed prairies.

Switch Grass
Panicum virgatum

Grass Family (Poaceae)

Description: Switch Grass is found in 2 large clumps on the Mound. It is 3 or 4 feet tall in the fall. Leaves are 18 inches long and 1/4 inch wide. The wispy seed heads are so small they are difficult to photograph. At the end of thin stems, the tops branch into many thin arms with very small purple flowers at the tips.

Location: Only recorded on the east side, about 10 yards from the east fence and 25 yards from the south fence, then again about 25 yards north.

Season: Emerging in April, flowering in September and October.

Comments: One of the four major prairie grasses that indicate a healthy prairie. Ours appears to be the Lowland Switch Grass, as it is a large, isolated clump and situated in a frequently wet drainage area. The *Panicum* genus has over 600 species, the largest genus of grasses.

Canada Wild Rye
Elymus canadensis

Grass Family (Poaceae):
Nodding Wild Rye

Description: Growing 3 to 5 feet tall on strong stems, this grass is easily identified by its 4-inch long wheat-like cluster of drooping flowers at the tips of the stem—green in the summer, then brown in the fall. The leaves are 1/4 inch wide and 15 inches long with sharp points.

Location: Scattered in small groups throughout the Mound.

Season: April to November.

Comments: The Indians are said to have used the seeds as food.

Chapter 9
The Dormant Season

With the ban on burning of the Mound in place, the next best treatment, mowing, is current policy. After the first freeze in November and before spring emerges, the Mound is closely mowed with a light tractor to mulch the bulk of the last year's growth. This treatment achieves some of the results of burning but is still only second best. It is hoped that someday a sympathetic town administration will find a way to again do what is best for the prairie.

The mulch will decompose over the winter and provide nutrients for new growth in February when the cycle of life on the Flower Mound begins again, as it has for the last ten thousand years or so.

The dormant season. The Mound in a rare snowfall. Winter, February 2004.

Mound Species Identified but Not Reproduced

Flowers

Fragrant Lily, *Androstephium caeruleum*
Henbit, *Lamium amplexicaule*
Japanese Honeysuckle, *Lonicera japonica*
Pepper Grass (Pepperweed), *Lepidium medium*
Bachelor's Button (Cornflower), *Centaurea cyanus*
Hooker's Plantain, *Plantago hookeriana*
Silverleaf Nightshade (White Horsenettle),
   Solanum elaeagnifolium
Pokeberry (Great Pokeweed), *Phytolacca americana*
Snow-on-the-Prairie, *Euphorbia bicolor*
Late-Flowering Boneset, *Eupatorium serotinum*
Sweet Scabious, *Scabiosa atropurpurea*
Texas Sage, *Salvia texana*

Grasses

Tall Dropseed, *Sporobolus asper*
Virginia Wild Rye, *Elymus virginicus*
Japanese Brome, *Bromus japonicus*
Wild Oats, *Avena fatua*
Johnson Grass, *Sorghum halepense*
Dallis Grass, *Paspalum dilatatum*
Italian Ryegrass, *Lolium perenne L. ssp. multiflorum*
Silver Bluestem, *Bothriochloa saccharoides*
Texas Bluegrass, *Poa arachnifera*
Winter Wheat, *Triticum aesticum*

Bibliography

Ajilvsgi, G. *Wildflowers of Texas*. Fredericksburg, Tex.: Shearer Publishing, 1984.

———. *Wildflowers of the Big Thicket, East Texas, and Western Louisiana*. College Station: Texas A&M University Press, 1979.

Bates, E. F. History and Reminiscences of Denton County. [Denton, Tex.]: Denton County Historical Commission, 1989.

Brown, Lauren. *Grasses, an Identification Guide*. Boston: Houghton Mifflin, 1979.

Collins, Scott L., and Linda L. Wallace. *Fire in North American Tallgrass Prairies*. Norman, [Okla.]: University of Oklahoma Press, 1990.

Denison, E. *Missouri Wildflowers: A Field Guide to Wildflowers of Missouri and Adjacent Areas*. Jefferson City, Mo.: Missouri Dept. of Conservation, 1989.

Diggs, G., B. Lipscomb, and R. O'Kennon. *Shinners & Mahler's Illustrated Flora of North Central Texas*. Fort Worth, Tex.: Botanical Research Institute of Texas and Austin College, 1999.

Elias, T. S., and P. A. Dykeman. *Edible Wild Plants: A North American Field Guide*. New York: Sterling Publishing Co., 1990.

Enquist, M. *Wildflowers of the Texas Hill Country*. Austin, Tex.: Lone Star Botanical, 1987.

Fernald, M. L., and A. C. Kinsey. *Edible Wild Plants of Eastern North America*. New York: Dover Publications, 1996.

Ferring, C. Reid. *An Archaeological Survey of Proposed Trails, Flower Mound, Texas*.

Foster, S. *A Field Guide to Medicinal Plants and Herbs*. Boston: Houghton Mifflin, 1999.

Freeman, C. C. *Roadside Wildflowers of the Southern Great Plains*. Lawrence, Kan.: University Press of Kansas, 1991.

Heatherley, A. N. *Healing Plants: A Medicinal Guide to Native North American Plants and Herbs*. New York: Lyons and Burford, 1998.

Kirkpatrick, Z. M. *Wildflowers of the Western Plains: A Field Guide*. Austin: University of Austin Press, 1992.

Ladd, D. *Tallgrass Prairie Wildflowers: A Falcon Field Guide*. Helena, Mont.: Falcon Press, 1995.

Lorrain, Paul, and Jan Lorrain. "Prehistoric Occupation of Lower Denton Creek." *The Record*, vol. 50 (September 2001).

Loughmiller, C., and L. Loughmiller. *Texas Wildflowers: A Field Guide*. Austin: University of Texas Press, 1999.

Matyas, M. "The Rise and Fall of Flower Mound New Town, an Interview with Marty Matyas."

McCormick, O. *Environmental, Floral, Faunal, and Ethnohistoric Study of Flower Mound New Town Area*. 1972.

Millspaugh, C. F. *American Medicinal Plants: An Illustrated and Descriptive Guide to Plants…Used in Medicine*. New York: Dover Publications, 1974.

Nasher, R. "Flower Mound New Town." Prospectus, Raymond Nasher Company, 1972.

Neihaus, T. F., C. L. Ripper, and V. Savage. *A Field Guide to Southwestern and Texas Wildflowers*. Boston: Houghton Mifflin, 1984.

Newcomb, W. W., Jr. *The Indians of Texas from Prehistoric to Modern Times*. Austin: University of Texas Press, 1961.

Niering, W., and N. Olmstead. *The Audubon Society Field Guide to North American Wildflowers, Eastern Region*. New York: Knopf, 1988.

Phillips, H. Wayne. *Plants of the Lewis and Clark Expedition*. Missoula, Mont.: Mountain Press Publishing Co., 2003.

Phillips, J. *Wild Edibles of Missouri*. [Jefferson City, Mo.]: Missouri Dept. of Conservation, 1979.

Runkel, S. T., and D. M. Roosa. *Wildflowers of the Tallgrass Prairie: The Upper Midwest*. Ames, Iowa: Iowa State University Press, 1989.

Saunders, Charles Francis. *Useful Wild Plants of the United States and Canada*. New York: Robert M. McBride & Co., 1934.

Smith, F. Todd. The Wichita Indians: Traders of Texas and the Southern Plains, 1540–1845. College Station: Texas A&M University Press, 2000.

Spellenberg, R. *The Audubon Society Field Guide to North American Wildflowers, Western Region*. New York: Knopf, 2001.

Vick, S. *Chronological History of the Ownership of the Mound*. Denton County Courthouse Records, Trax Company, 2002.

Whetsell, W. C. *Pasture and Range Plants*. Hays, Kan.: Fort Hays State University, 1989.

Wills, M. M., and H. S. Irwin. *Roadside Flowers of Texas*. Austin: University of Texas Press, 1961.

Index

A
Acacia angustissima, 80
Achillea millefolium, 55
Agalinis purpurea, 95
Alexander, Clementine, 26
Alexander, James, 26
Allium drummondii, 54
Ambrosia trifida, 96
Andropogon gerardii, 101
Androstephium caeruleum, 108
Anemone berlandieri, 36
Antelope Horns, 75
Arnoglossum plantagineum, 61
Asclepias viridiflora, 84
Asclepias viridis, 75
Aster ericoides, 95
Astragalus crassicarpus, 39
Avena fatua, 108

B
Bachelor's Button, 108
Baldwin, William, 93
Basket Flower, 84
Beargrass, 60
Beavers, George and Hattie, 19
Bellamah Community, 22
Benson, Henry "Chief," 30
Bettinger, Edith, 7, 29
Big Bend, 45
Big Bluestem, 101
Bindweed, 83
Black Eyed Susan, 77
Black Land Prairie, 9, 11
Black Mark Farms, 4, 5, 20, 21, 22
Bladderpod, 37
Blazing Star, 98
Blue Sage, 71
Bluebonnet, Texas, 45
Blue-eyed Grass, 51
Bluet, 50
Boston Tea Party, 62
Botanical Research Institute, 7
Bouteloua curtipendula, 104
Bowman, Alton, 5, 8, 32, 112
Bowman, Sweety, 32, 33
Bromus japonicus, 108
Broomweed, Common, 92
Brown Eyed Susan, 77
Buffaloweed, 96

C
Callirhoe alcaeoides, 43
Camassia scilloides, 48
Campbel, Ben, 29
Canada Wild Rye, 106
Carduus nutans, 89
Carter, Gregory S., 29
Castilleja indivisa, 58
Celestial, 46
Centaurea americana, 84
Centaurea cyanus, 108
Cherokee Indians, 66, 77
Cirsium texanum, 86
Clasping Leaved Coneflower, 82
Clover, White Prairie, 91

Clover, Yellow Sweet, 66
Cobo, Father Bernardo, 57
Common Bindweed, 83
Common Broomweed, 92
Common Dandelion, 41
Common Sunflower, 79
Common Vetch, 76
Compass Plant, 89
Coneflower, Clasping Leaved, 82
Consolvo, Otto and Babe, 26, 29
Convolvulus equitans, 83
Cooperia drummondii, 78
Cornflower, 108
Crawford, Jerry John, 29
Cristal, John and Richard, 15
Crow Poison, 38
Cut Leaf Daisy, 69

D
Daisy, Engelmann, 69
Dalea multiflora, 91
Dallas Museum of Natural History, 14
Dandelion, Common, 41
Dandelion, Texas, 44
Death Camas, 48
Delphinium virescens, 63
Dogan, Nicole, 7, 32
Donald Academy, 25
Donald Community, 25
Donald, Matthew Lyle, 26
Dracopis amplexicaulis, 82
Dyschoriste linearis, 74

E
Eads, Andrew, 8, 32
Eastern Cross Timbers, 9, 11
Eidson, Jim, 7
Elymus canadensis, 106
Engelmann Daisy, 69
Engelmann, Dr. George, 69
Engelmannia pinnatifida, 69
Erigeron philadelphicus, 50
Evening Primrose, 44
Evening Star, 78

F
False Garlic, 38
Faulconer, Rosa Lee, 31
Fern Acacia, 80
Field Pansy, 80
Firewheel, 85
Fleabane, Philadelphia, 50
Flower Mound Fire Department, 31
Flower Mound New Town, 22
Flower Mound Presbyterian Church, 26
Foxglove, Wild, 57
Fragrant Lily, 108
French, Willard, 29
Fringed Puccoon, 42

G
Gaillardia pulchella, 85
Gaura coccinea, 64
Gayfeather, 98
Gerault, Andre, 32
Gerault, Levenia, 32
Geum canadense, 77

Gilgai (hog wallows), 33
Glandularia bipinnatifida, 47
Goatsbeard, 70
Goldenrod, 97
Goldenweed, 93
Grapevine Lake, 102
Great Pokeweed, 108
Green Antelope Horns, 84
Green Flower Milkweed, 84
Green Milkweed, 75
Greene, A. C., 7, 26
Grindelia papposa, 93
Ground Plum, 39
Groundsel, 40
Gutierrezia dracunculoides, 92

H
Hanson, JoAnn, 8
Hartwell, Scott, 8
Heath Aster, 95
Hedyotis nigricans, 50
Helianthus annuus, 79
Helianthus maximiliani, 92
Henbit, 108
Heslinga, Betsy, 31
Hilliard, Ernest, 29
Hilliard, Ronnie, 29, 30, 31
Holland Coffee Company, 16
Hooker's Plantain, 108
Horsenettle, Western, 67
Horsenettle, White, 67, 108
Hyacinth, Wild, 48
Hymenopappus scabiosaeus, 65

I
Immel, John, 31
Indian Blanket, 85
Indian Grass, 103
Indian Paintbrush, 58
Indian Plantain, 61
Italian Ryegrass, 108

J
Japanese Brome, 108
Japanese Honeysuckle, 108
Johnny-Jump-Up, 80
Johnson, Lady Bird, 4, 68

K
Kansas, 79
Krameria, 49
Krameria lanceolata, 49

L
Lactuca serriola, 89
Lake Forest Development, 29, 31
Lamium amplexicaule, 108
Lance Leaf Loosestrife, 78
Larkspur, Prairie, 63
LBJ National Grasslands, 48
Leone, Anita, 32
Lepidium medium, 108
Lesquerella gordonii, 37
Lester, Ben, 27
Lester, Ray and Charlie Fay, 20
Lewis and Clark, 60
Liatris mucronata, 98
Lily, Fragrant, 108

110

Lily, Rain, 78
Lindheimer, Ferdinand, 47
Lindheimera texana, 47
Lindheimer's Daisy, 47
Lithospermum incisum, 42
Little Bluestem, 102
Locoweed, 68
Lolium perenne L. ssp. multiflorum, 108
Lonicera japonica, 108
Louisiana Purchase, 16
Low Ruellia, 90
Lowland Switch Grass, 105
Lupinus texensis, 45
Lythrum alatum, 78

M
Marcus, Betty, 20
Marcus, Edward, 4, 5, 20, 30, 68
Matyas, Marty, 7, 22
Maximilian Sunflower, 92
McCarley, Bob and Florence, 27
Medlin, Hall, 15
Melilotus indicus, 66
Mexican Hat, 51
Milkweed, Green, 75
Milkweed, Green Flower, 84
Mimosa roemeriana, 59
Mirabilis linearis, 94
Monarda citriodora, 62
Musk Thistle, 89

N
Narrow Leaf Gayfeather, 98
Nasher, Raymond, 22
Nassella leucotricha, 104
Neiman, Bill, 7, 31
Nelson, Joan, 8, 32
Nelson, John, 32
Nemastylis geminiflora, 46
Nodding Thistle, 89
Northoscordum bivalve, 38

O
Oenothera speciosa, 44
Ohio Spiderwort, 56
Oklahoma, 85, 103
Old Plainsman, 65
Oxytropis lambertii, 68

P
Panicum virgatum, 105
Penstemon cobaea, 57
Pepper Grass, 108
Pepperweed, 108
Peters Colony, 16
Pettyjohn, Barbara, 31
Philadelphia Fleabane, 50
Phlox pilosa, 56
Physalis pubescens, 72
Phytolacca americana, 108
Picardi, Albert, 8, 32
Pink Prairie Rose, 73
Plains Beebalm, 62
Plains Winecup, 43
Plaintain, Indian, 61
Plantago hookeriana, 108
Plantain, Hooker's, 108
Poa arachnifera, 108
Poison Ivy, 74
Pokeberry, 108
Polytaenia nuttallii, 65
Poppy Mallow, 43

Popweed, 37
Potato Weed, 67
Prairie Acacia, 80
Prairie Larkspur, 63
Prairie Onion, 54
Prairie Parsley, 65
Prairie Phlox, 56
prairie plow, 17
Prairie Verbena, 47
Preston Bend, 16
Prickly Lettuce, 89
Primrose, Evening, 44
Prince Maximilian, 92
Purple Gerardia, 95
Purple Horsemint, 62
Pyrrhopappus multicaulis, 44

Q
Queen Anne's Lace, 55

R
Ragweed, 96
Ragwort, 40
Rain Lily, 78
Ratany, 49
Ratibida columnaris, 51
Rheudasil, Bob, 4, 7, 20, 22, 30
Ringmacher, Nelson, 8, 32
Robinson, Forest, 31
Rosa foliolosa, 81
Rosa setigera, 73
Rose, Pink Prairie, 73
Rose, White Prairie, 81
Rubus trivialis, 41
Rudbeckia hirta, 77
Ruellia humilis, 90

S
Salvia azurea, 71
Salvia texana, 108
Saw-leaf Daisy, 93
Scarlet Gaura, 64
Schizachyrium scoparium, 102
Senecio ampullaceus, 40
Sensitive Briar, 59
Shannon, Ralph, 31
Showy Primrose, 44
Sideoats Grama, 104
Silverleaf Nightshade, 67, 108
Sisyrinchium pruinosum, 51
Smith, Doris, 31
Snake Herb, 74
Solanum dimidiatum, 67
Solanum elaeagnifolium, 67, 108
Solidago altissima, 97
Sonchus asper, 98
Sorghastrum nutans, 103
Southern Dewberry, 41
Sow Thistle, 98
Sporobolus asper, 108
Staten Oak, 13
Sternberg Museum, 17
Summit Club, 25
Sunflower, Common, 79
Sunflower, Maximilian, 92
Swenson, John, 32
Switch Grass, 105

T
Tall Dropseed, 108
Taraxacum officinale, 41
Ten-Petal Anemone, 36

Teucrium canadense, 90
Texas Archaeological Research Laboratory, 33
Texas Bindweed, 83
Texas Bluebonnet, 45
Texas Bluegrass, 108
Texas Dandelion, 44
Texas Nature Conservancy, 7
Texas Sage, 108
Texas Star, 47
Texas State Historical Commission, 30
Texas Thistle, 86
Texas Vervain, 64
Texas Wintergrass, 104
Texas Yellow Star, 47
Thimble Flower, 51
Thistle, Musk, 89
Thistle, Texas, 86
Thompson, Bruce, 8, 32
Topographical Map, 9
Toxicodendron radicans, 74
Tradescantia ohiensis, 56
Tragopogon dubius, 70
Trammel Crow Company, 30
Triodanis perfoliata, 66
Triticum aestivum, 108

V
Valerianella woodsiana, 76
Venus' Looking-glass, 66
Verbena halei, 64
Verbena, Prairie, 47
Vernonia baldwinii, 93
Vervain, Texas, 64
Vicia sativa, 76
Viola rafinesquii, 80

W
Webb, Mary, 29
Western Horsenettle, 67
Western Ironweed, 93
White Avens, 77
White Prairie Clover, 91
White Prairie Rose, 81
Wichita Indians, 14, 15, 16
Wild Bergamot, 62
Wild Dill, 65
Wild Four O'Clock, 94
Wild Foxglove, 57
Wild Garlic, 54
Wild Hyacinth, 48
Wild Oats, 108
Wild Petunia, 90
Wild rose hips, 73
Wilkins, Dave, 32
Wilkins, Lori, 8, 32
Winter Wheat, 108
Wise, Arlene, 29
Wise, Jack, 29, 32
Wiswell, John and Edy, 19
Women of Flower Mound, 25
Wood Sage, 90
Woodbine Formation, 11
Woods Corn Salad, 76

Y
Yarrow, 55
Yellow Ground Cherry, 72
Yellow Sweet Clover, 66
Yucca, 60
Yucca arkansana, 60

About the Author

Alton Bowman was born in Vincennes, Indiana, on June 19, 1945. He has resided in Flower Mound since 1984 with his wife, Sweety, and their three children, Ada, Alton IV, and Ariel. Mr. Bowman is a furniture conservator by profession and has restored important works such as the Supreme Court bench for the Texas State Capitol, the Rosedown bed for the Dallas Museum of Art, and the furnishings of the Moody Mansion in Galveston, as well as items in many public and private collections. Mr. Bowman served as the first chairman of the Flower Mound Tree Board, which was instrumental in developing the Flower Mound Tree Ordinance and the town's Tree City USA designation. He has served on the Mound Foundation since 1987 and has been its chairman since 1993.